DIALOGUES ON CONSCIOUSNESS

DIALOGUES ON CONSCIOUSNESS

Riccardo Manzotti and Tim Parks

OR Books
New York · London

All rights information: rights@orbooks.com
Visit our website at www.orbooks.com
First printing 2020

Library of Congress Cataloging-in-Publication Data: A catalog record for
this book is available from the Library of Congress.

Typeset by Lapiz Digital Services.

paperback ISBN 978-1-68219-224-5 • ebook ISBN 978-1-68219-226-9

CONTENTS

1.	The Challenge of Consciousness	1
2.	The Color of Consciousness	11
3.	Does Information Smell?	23
4.	The Ice Cream Problem	35
5.	Am I the Apple?	49
6.	The Mind in the Whirlwind	63
7.	Dreaming Outside Our Heads	75
8.	The Body and Us	89
9.	Consciousness: Who's At the Wheel?	101
10.	A Test for Consciousness	113
11.	The Hardening of Consciousness	127
12.	Consciousness: An Object Lesson	139
13.	The Pizza Thought Experiment	151
14.	Consciousness: Where are Words?	163
15.	Consciousness and the World	175

1. THE CHALLENGE OF CONSCIOUSNESS

Is it possible to put some order into our thoughts about consciousness, memory, perception, and the like? Hardly a day goes by without some in-depth article wondering whether computers can be conscious, whether our universe is some kind of simulation, whether the mind is a unique quality of human beings or spread out across the universe like butter on bread. Many of us are not even sure what we believe in this department or whether what we believe would bear much scrutiny from philosophers or neuroscientists.

For a number of years I have been talking about these matters almost daily with Riccardo Manzotti, the philosopher, psychologist, and robotics engineer. I have now suggested to him that we condense our conversations into a series of focused dialogues to set out the standard positions on consciousness, and suggest some alternatives. For my own part, I'd like to add some reflections on the social implications of the various theories for what we think about consciousness, which is as much as to say what we think about who and what we are inevitably has consequences for how we relate to one another, and to the world. But our first problem will be one of definition.

—Tim Parks

Tim Parks: Riccardo, what do we mean when we say "consciousness"? Are we talking about perceptive experience, memory, thought, trains of thought, or mental life in general?

Riccardo Manzotti: For most people "consciousness" will have various meanings and include awareness, self-awareness, thinking in language. But for philosophers and neuroscientists the crucial meaning is that of feeling something, having a feeling you might say, or an experience. An easy way to think about it would be pain. Instinctively we all agree that feeling a pain is something. It's an experience. That is why we don't like to hurt animals, because we have good reason to suspect that they feel what happens to them. And this feeling of what happens to us characterizes our existence. The technical term is "phenomenal experience," or again "conscious experience," but frankly both sound a tad redundant since experience is always something we feel.

Parks: I remember David Chalmers, a philosopher we'll no doubt be talking about at some point, defining consciousness as an internal flow of images, "a movie playing inside your head," and probably a lot of people would agree with him. But you want to stick to something more basic.

Manzotti: A definition like that suggests that we know a lot more than we do: that there are images in our heads, that they move forward in sequence, that there is some kind of split between the image and someone (who?) observing the image. It's all very problematic. The truth is that we do not know what consciousness is.

That's why we're talking about it as a problem. What we do know is that the way we experience reality, the way we feel the things that happen to us, does not really match up with our current scientific picture of the physical world.

Parks: In what respect?

Manzotti: Well, consider this: If we didn't know that human beings experience the world, that they feel things, would we be able to deduce it from what we know about neurophysiology? Really, no. There is nothing about the behavior of neurons to suggest that they are any different with respect to consciousness than, say, liver cells or red blood cells. They are cells doing what cells do best, namely, keeping entropy low by generating flows of ions such as sodium, potassium, chloride, and calcium, and releasing neurotransmitters as a consequence. All of that is wonderful but far removed from the fact that I experience a light blue color when I watch the morning sky. That is, it's not easy to see how the physical activity of the neurons explains my experience of the sky, let alone a process like thinking.

Parks: So we might say that consciousness is the word we use to refer to the fact that rather than just physiological activity, mute like any other physical event—the sky in the morning, a cloud crossing the sun—we have experience, we have a feeling of that event?

Manzotti: Exactly. Instead of a world where we merely interact with external occurrences—the way a flower opens in the sun, or water freezes in the cold—we also have experience of the occurrence: the sun, the icy weather, and so on. This addition of experience (or in future we may want to suggest that experience and occurrence are one!) would be puzzling enough in itself. But it is even more puzzling that experience is usually described as experience *of something else*, of something that is not me. I experience a red apple. You experience a piece of music. Ruth experiences a landscape. How is this possible since, if we leave aside quantum mechanics (for the moment), our traditional view of nature tells us that an object is what it is and nothing more? William James put this very clearly when he asked, "How can the room I am sitting in be simultaneously out there and, as it were, inside my head, my experience?" We still have no answer to that question.

Parks: So another way we could look at this would be to say that the fact of consciousness points to a flaw in our explanation of reality. Or at least amounts to a big challenge as to how we understand reality.

Manzotti: Right. Once we have defined and placed all the pieces of the physical jigsaw—chemistry, physics, evolution, general relativity, quantum mechanics, DNA, evolution, Higgs boson, the lot—there is still something that does not add up—namely the fact that we don't simply *do* things, we also *experience* the world around us. Consciousness. What David Chalmers famously called the hard problem.

Parks: In other words, consciousness is not something that current science would predict.

Manzotti: No. Why doesn't our behavior simply happen, taking its course the way the planets follow their orbits? We don't know. Just as cosmologists don't know what dark matter is. All we know is that there is something that doesn't add up and very likely points to some profound error in our assumptions about reality. That's what we should be concentrating on, rather than getting into elaborate and suggestive metaphors like "movies in the head."

Parks: You seem now to be defining consciousness by what it is not, or at least as an area of incomprehension. But can I push you toward a more positive definition? I mean, are we talking about a thing—a physical object or a process? I presume we rule out spirits and souls.

Manzotti: To speak of spirits and souls would amount to an admission of defeat, at least for a scientist or philosopher. The truth is that we just don't know *a priori* the nature of physical reality. This is a point Bertrand Russell made very strongly back in the 1920s. The more we investigate the physical, the more varied and complex it appears. Imagine a huge puzzle in which everything must fit together with everything else. When there's something that doesn't seem to match up, we turn it this way and that to see if we can make it fit somehow, but if it won't, we have to assume that we've put the other pieces together wrongly: we've got a false picture.

   That's how science proceeds. So we have moments of revolution—Copernicus, Galileo, Newton—when all the pieces have to be rearranged, what Thomas Kuhn famously described as paradigm shifts.

There's no reason why we should approach the problem of consciousness any differently. We have to find how to fit it into our existing understanding of reality, or change our version of reality to have it fit in with consciousness. Until we do that, we risk having a dualistic vision of the world, like the one suggested by Descartes: on the one side the physical, on the other something rather mysterious, call it the spiritual.

Parks: But again, should we be thinking of consciousness as a thing, or a process?

Manzotti: Well, if the world that surrounds us is made of things, objects, and physical processes, consciousness is likely to be one of them. People tend to be extremely hesitant when approaching consciousness and treat it as a special case. But I'm not sure that's helpful. If it is a real phenomenon, and most people agree that it is, why shouldn't it be like all other physical phenomena, something made of matter and energy whose activity is explicable by its physical properties?

Parks: So, assuming consciousness is a thing, a physical thing—or an amalgam of things—what do we do with the word "mental"?

Manzotti: Good question! Actually "mental" isn't so different, at least as regards its function, from a word like "spiritual." Neither word has a precise referent. I'm afraid we're going to run into a lot of words like this in the course of these conversations. It's as if certain terms we use had been given a special license to operate outside the constraints of the physical world. The philosopher Sidney Shoemaker observed that the notion of the "mental" amounts to a kind of ontological dustbin. Anything that doesn't fit with our current picture of physical reality is moved to the bin whose main purpose is to collect together all the things we can't explain. It's a sort of quiet dualism: you don't say the word "spirit," but in fact you're splitting the world in two.

Parks: A bin is hardly flattering. Surely when we talk about our mental lives we're simply thinking of everything that makes human beings special, different—our thoughts, our language-based lucubration.

Manzotti: Absolutely. There are good reasons to be fond of a notion like "the mental," because it places our minds above the constraints of physical necessity. It's a comforting idea. We are above nature. We are special. We have our mental lives. Separate from the nitty-gritty of matter. Unfortunately, we have no scientific justification for this belief, which is very likely just another manifestation of what Freud described as human narcissism, the desire to believe ourselves at once at the center of the universe, yet in some way superior to and even separate from the nature around us.

How convenient, when you can't explain something, to say, well, that means we're special, we're not like the rest of the natural world. But science works on the assumption that nature is one and that all phenomena must fit in the same system and obey the same laws; hence the fact that we experience the world—i.e., consciousness—must be a natural phenomenon which, like all other natural phenomena, is physical, made of matter and energy.

Parks: This brings us, I think, to the dominant view of what consciousness is today: internalism. Can you explain?

Manzotti: Internalism is the notion that whatever consciousness is, it must happen inside the head. It's fairly obvious why we might think this. We tend to feel that we are located where our senses are; hence people suppose that consciousness is somewhere behind our eyes and between our ears. This not to mention the many social reasons for identifying with our bodies in general and our faces in particular, which are crucial to social interaction. And since of course we can't *see* consciousness in another person, but only manifestations of it—smiles, grimaces—we assume it is hidden inside the head, that is, in the brain. Since, again, the brain is by far the most complex of our organs, with something like 85 billion neurons, all with hundreds if not thousands of connections to other neurons, it seems a reasonable candidate when you're looking for something you don't understand. Or it did seem so when we knew less about it.

Parks: I know you have strong objections to internalism and can feel you straining at the leash to express them. But let's first establish exactly what the theory is and what it claims. For example, does internalism claim that consciousness is a physical object located in space?

Manzotti: There are many strands to internalism, but on the whole, and certainly initially, yes. The idea was formalized in the 1950s by people like D.M. Armstrong and J.J.C. Smart. They advanced the idea that consciousness *is* neural processes, or certain neural processes. Once they'd formed this perfectly respectable hypothesis an army of scientists set about verifying it empirically. And in fact, over the past fifty years we've made extraordinary progress in the development of sophisticated instruments to probe and explore the brain with all its fantastically intricate electrical and chemical activity.

Parks: And?

Manzotti: Well, neuroscientists have certainly found a huge number of *correlates of consciousness*; that is, for all kinds of sensory experiences they have established which parts of the brain are active, and the nature of that activity. This is of enormous interest and scientifically very sound.

Parks: I hear a "but" coming.

Manzotti: Well, a *correlate* of consciousness is not consciousness. When scientists look for AIDS or DNA, they look for the thing itself, not a mere correlate. This is a problem: how to get from the neural correlate—the fact that there's neural activity when

I experience something—to the thing itself, the experience? As Thomas Nagel almost facetiously put it, when one licks chocolate ice cream nothing in the brain tastes like chocolate. Of course, an experience also has correlates outside the brain: the sensory organs—eyes, ears, nose, skin, taste buds—not to mention the object itself that we experience: light, soundwaves, that chocolate ice cream, whatever. Why privilege the correlates in the brain in our attempt to locate consciousness? Why...

Parks: Stop there! Enough for today. We've defined consciousness as the feeling that accompanies our being in the world. We've looked very crudely at the conundrum its existence poses for our understanding of the world. We've announced the dominant scientific view of where consciousness is located: in our brains. Next time, I'd like to consider some of the claims of internalism, their implications for our current scientific account of reality, and the way internalists have reacted to their difficulties verifying their theory. Because they certainly haven't given up. Far from it. So be prepared.

2. THE COLOR OF CONSCIOUSNESS

There are no colors out there in the world, Galileo tells us. They only exist in our heads. In the first of our dialogues about the mind, Riccardo Manzotti and I established that by "consciousness" we mean the feeling that accompanies our being alive, the fact that we experience the world rather than simply interacting with it mechanically. We also touched on the problem that traditional science cannot explain this fact and does not include it in its account of reality. That said, there is a dominant understanding of where consciousness happens: in the brain. This "internalist," or inside-the-head, approach shares Galileo's view that color, smell, and sound do not exist in the outside world but only in the brain. "If you could perceive reality as it really is," says leading neuroscientist David Eagleman, "you would be shocked by its colorless, odorless, tasteless silence." What Riccardo and I want to do today is ask how, in the neuroscientists' opinion, we see color. What are the implications of believing that this experience is all inside our heads? And how have scientists reacted to the difficulties they have encountered verifying this theory?

—Tim Parks

Tim Parks: Riccardo, when the internalists talk about conscious experience, they often use the word "qualia," meaning an elementary sensation, a feeling of something, and one of their favorite examples of this is our seeing color, our experience of color. So how does it come about that we see color?

Riccardo Manzotti: Before answering let's pay some attention to the language we're using, since it may determine the way we think about the whole thing. Most people say they see a color or a colored object, a yellow banana, say. So we have subject and object; a person sees a yellow banana. Scientists and philosophers speak of our having an experience, feeling, or qualia. So now we have three things, a subject, an object (the banana maybe), *and* a feeling, in our heads. I fear both manners of speaking are potentially misleading.

Parks: I suppose it's inevitable that standard views of experience will be built into language use, but can't we leave this issue for another time?

Manzotti: I'm not sure we can. The fact is that the subject/object divide, not to mention the addition of a feeling or "percept," is particularly pertinent when we talk about color.

Parks: How so?

Manzotti: Well, as you said, science tells us there's no color in the world. It occurs only in our brains. But, as we discussed in our first conversation, when scientists look inside the brain to see what's going on, they find only billions of neurons exchanging electrical impulses and releasing chemical substances. They find what they call correlates of consciousness, not consciousness itself; or in this case, they find correlates of color, but not color itself. There is no yellow banana in the head, just the grey stuff.

Parks: I get it. We have no color outside in the world and yet we can't find the color in our heads either. So where on earth is it? The funny thing is that for most people there is no problem at all. They see a red traffic light and they know to stop. They trust other people to stop. Most of us feel entirely confident about seeing color and even mixing colors. To us color seems to be an external reality, not a subjective delusion.

Manzotti: Absolutely! The unsuspecting layman will assure you that objects simply *have* colors as attributes—isn't our banana, for example, very yellow exactly the same way it's six inches long? Well, unfortunately not, because it could easily be shown that bananas are only yellow *under a certain light*. Change the light and the banana might look green. But it will always be six inches long.

Parks: I had thought color was revealed, as it were, by a refraction, or breaking up of light.

Manzotti: That is the other traditional claim, still widely taught in school, that colors exist *in light*, or that different colors are different *wavelengths of light*. And of course the colors of the rainbow immediately come to mind. But that explanation doesn't work 100 percent either. The same wavelength, for example, will give rise to different colors if the surrounding environment is different. To his credit, Newton himself, who actually introduced the word "spectrum" into the English language to refer to the range of possible colors, eventually dismissed the idea that colors are literally contained in the light. "For the Rays, to speak properly, are not coloured. In them there is nothing else than a certain Power and Disposition to stir up a Sensation of this or that Colour." Three hundred years on, what and where colors actually *are* remains a mystery.

Parks: Yet so many books on neuroscience purport to tell us how we perceive color.

Manzotti: Indeed. From the nineteenth century on scientists have been looking for colors inside the nervous system. First, they worked on the hypothesis that colors were qualities that the retina or the optic nerve introduced into the signals that they then sent to areas further along in the brain. However, nothing satisfactory was ever found, nor is it clear what they imagined they might find. In the twentieth century neurophysiologists went deeper and deeper into the brain, tracing the activity of neurons they believed related to the experience of color, until, around 1973, a provisional consensus was reached when the neuroscientist Semir Zeki presented evidence that part of the visual cortex in the occipital lobe at the back of the head, an area called V4, was responsible for color perception, this because damage to that area led to a loss of color perception and color memory. But once again, further research suggested that matters were more complicated, that other areas and neurons could come into play. In short, there is still a great deal of debate on the subject.

Parks: I appreciate that you've spent a great deal of time research-ing the history of science's dealings with color, but are you telling me that contemporary neuroscience offers no dominant view on the matter?

Manzotti: Well, the current textbook view goes like this. The world is a place where objects reflect light, sunlight being the dominant source and as it were, the default setting as far as the kind of light is concerned. However, each object reflects only a subset of that light. Rays from this subset enter our retina and stimulate a honeycomb of cells, known as cones, because of their conical shape, whose function is to react differently to different portions of the visible spectrum (we remember, of course, that only a small part of the vast electromagnetic spectrum is visible). Most humans—animals are rather different—possess three kinds of cones, referred to as S, M, and L cones, depending on whether they react more vigorously to short, medium, or long-range light wavelengths. The "output" of these cells is first merged together in the retina, then sent via the optical nerve to various cortical areas—including the famous V4. And that's as much as we know.

Parks: Riccardo, you just gave me the whole explanation without ever using the word color.

Manzotti: I know! Oddly, this is a theory of color that does not need the notion of colors. I suppose the reason is that however carefully you follow neural signals from the retina along the optic nerve and across the brain, you don't actually come across anything like a color, or anything that explains color perception. You could almost say that the notion of color is useless to color science, unless...

Parks: Unless?

Manzotti: Unless we bring consciousness back in the picture. Colors are something we *experience,* individually and collectively. But without our experience of color, science would have no reason to suspect its existence. There would just be fifty shades— or more likely fifty thousand shades—of electromagnetic waves. That is why even a Nobel Prize-winning biologist like Gerald Edelman tells us that reality is actually colorless; because he takes reality to be what science tells us it is, not what he experiences as an individual.

Parks: But the implications of this "official" view are profound. First, it suggests our perceptions are radically separate from the external world, fenced off inside the skull. Second, and as a result, that we all live in error and need the authority of science to tell us what reality is really like. So it gives scientists considerable power.

Manzotti: It's obvious that for modern science to happen, the object had to be separated from the subject; only an elite of savants could be acquainted with the thing itself. Remember that Galileo, the founder of modern science, was also a Platonist, and Plato was a prominent member of his city's elite, and the first philosopher to place the object of intellectual enquiry outside of the reach of the everyday man: that is, man is trapped in the cave watching shadows on the wall while reality is outside, beyond his grasp.

Parks: I think the thing that most disturbed me when I read Galileo was his exhortation, which Francis Bacon agreed with, that we should do violence to our senses, deliberately go against what they tell us.

Manzotti: Absolutely. *Far la ragione tanta violenza al senso.* If science tells you the world has no color, then you must fight against your perception that in fact it is leaping with color.

Parks: So, while in our first dialogue, you suggested that consciousness is a challenge for our present scientific model of reality, in that nothing in science predicts the existence of consciousness, now it seems you are going a step further and suggesting that consciousness is a kind of battleground between science and the lay community—with science telling us that conscious experience, color for example, is a brain-based illusion, while for the layman it is reality itself.

Manzotti: Right. You could say that going right back to its origins, science has struggled to keep the lay observer at bay, since his experience—in this case of color—stains the purity of science's mathematical description of reality (remember that for Plato, and indeed for Newton, a natural law was a divine law). "Our perception of reality," says Eagleman, "has less to do with what's happening out there, and more to do with what's happening inside our brain." In short, our experience is a kind of hallucination. The world is colorless. The yellowness of your banana is a fantasy. More than that, you...

Parks: Hold on! I know it's my fault for raising the question of social implications, but I fear we're losing focus. Can we try and stay on color and the internalist view of color. Can we sum up? Or did I make a big mistake choosing color as a starting point?

Manzotti: Well, you certainly chose something that brings to the fore the chasm between the neuroscientific, internalist account of consciousness, and our day-to-day experience. Let me try and clinch this with something called Kitaoka's illusion. Akiyoshi Kitaoka is a biologist and psychologist who has created all kinds of optical "illusions" to demonstrate our supposed fallibility—to catch our brains out, as it were. Check out the spirals here.

Kitaoka's illusion.

Looking at this, we see blue, green, and pink spirals. However, Kitaoka maintains that the two spirals that seem to be green and blue are in fact the same, because the pixels between the thin stripes in each spiral are the same. So the fact that we see one as green and the other as blue is an illusion.

Parks: I'm sorry, I really don't understand.

Manzotti: It's simpler than it seems. In computer science, engineers have devised a way to produce colors mixing three basic wavelengths, corresponding to red, green, and blue, in dots called pixels. They describe the colors in terms of a so-called RGB triplet, three numbers that specify the quantities of each wavelength in the mix. It is a crude but efficient recipe that allows you to obtain a huge number of colors (on your laptop screen, for example). However, you really can't predict the color you'll see from a single RGB triplet any more than you can from a light wavelength, since, in different environments, those triplets will lead to our experiencing different colors, as here with Kitaoka's spirals. If you check the RGB values for both the green and the blue spirals, you'll find they are indeed the same (0, 255, 150); what makes the difference is the color of the stripes across the spirals (orange where the spiral is green, magenta where it's blue).

Kitaoka, who's aiming to demonstrate a distance between appearance and reality, says that since the RGB triplets in the two spirals are the same, we should experience them as the same color.

And if we don't, it's because we're victims of an illusion. Let me draw you a cruder version on the iPad [below].

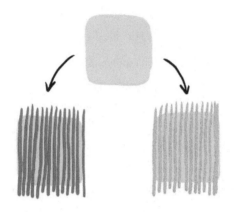

Manzotti explains Kitaoka's illusion.

Parks: Fascinating.

Manzotti: But hardly instructive! Anyone can see that there are blue spirals and green spirals. And everyone *does* see that, or at least everyone with what we call normal sight. Rather than insisting that the same pixels should always appear as the same color and talking about an appearance/reality gap, what science should be doing is accepting the reality of conscious experience and questioning its own obviously imperfect understanding of what constitutes color.

Parks: I recall now how the neuroscientist Christof Koch claims that our experience of color is "a con job." David Eagleman talks about vision being the result of "fancy editing tricks." The philosopher Eric Schwitzgebel accuses us of being "ignorant and prone to error" about everything we see. A moral nuance is smuggled into the debate, as if there were something shabby and lazy in the way we see the world. It feels like we're being blamed, or condescended to, for not perceiving things as science thinks we ought to.

Manzotti: Exactly. Kitaoka's illusion—and there are many others like it—becomes a flash point where the assertions of science in the Galileo tradition clash with the validity of our conscious perception.

Parks: So the question is, who is in denial? Science with an inadequate understanding of color, or the ordinary individual with his or her experience of it?

Manzotti: Right, so maybe what we need to do is to get beyond the idea that consciousness is a "representation" of the world at all. Maybe it is simply reality. Maybe, as I hinted at the beginning, we have to do away with that subject/object distinction which lies behind this whole discussion.

Parks: Maybe we do, but not today. I had promised that we would look at how internalists have responded to their failure, at least so far, to show how the brain produces consciousness, how that colorless world is colored in. But I now see that was a bridge too far. We'll leave it to the next dialogue, when the question will be this: How are internalists defending or proposing to defend a position that you obviously reckon is indefensible?

3. DOES INFORMATION SMELL?

In our first two dialogues, we presented the standard, or "internalist" version of how our conscious experience of the world comes about: very bluntly, it assumes that the brain receives "inputs" from the sense organs—eyes, ears, nose, etc.—and transforms them into the physical phenomenon we know as consciousness, perhaps the single most important phenomenon of our lives. We also pointed out, particularly with reference to color perception, how difficult it has been for scientists to demonstrate how, or even whether, this really happens. Neuroscientists can correlate activity in the brain with specific kinds of experience, but they cannot say this activity is the experience. In fact, the neural activity relating to one experience often seems nearly indistinguishable from the neural activity relating to another quite different experience. So we remain unsure where or how consciousness happens. All the same, the internalist model remains dominant and continues to be taught in textbooks and broadcast to a wider public in TV documentaries and popular non-fiction books. So our questions today are: Why this apparent consensus in the absence of convincing evidence? And what new ideas are internalists exploring to advance the science?

—Tim Parks

Tim Parks: Riccardo, I know I should be asking the questions, not answering them. But I'm going to suggest that one reason for this consensus is that we are in thrall to the analogy of the brain as computer. For example, a recent paper I was reading about the neural activity that correlates with the sense of smell begins, "The lateral entorhinal cortex (LEC) computes and transfers olfactory information from the olfactory bulb to the hippocampus." Words like "input," "output," "code," "encoding," and "decoding" abound. It all sounds so familiar, as if we knew exactly what was going on.

Riccardo Manzotti: We must distinguish between internalism as an approach to the problem of consciousness (the idea that it is entirely produced in the head) and neuroscience as a discipline. The neuroscientists have made huge progress in mapping out the brain and analyzing the nitty-gritty of what goes on there: which neurons are firing impulses in which rhythms to which others, what chemical exchanges are involved, and so on. But you are right: the way they describe their experiments by way of a computer analogy—in particular of information processing and memory storage—can give the mistaken impression that they're getting nearer to understanding what consciousness is.

When physiologists address other parts of the body—the immune system, the kidneys, our blood circulation—they don't feel the need to use anything but the language of biology. Read a paper on, say, the liver, and it will be talking about biochemical mechanisms—metabolites, ion homeostasis, acetaminophen poisoning, sepsis, infection, fibrosis, and the like, all terms that refer to actual physical circumstances. Yet, when dealing with the brain, we suddenly find that neurons are processing "information," rather than chemicals.

Parks: Is this because while we know what other organs are doing—I mean, which physical processes in the body each is responsible for—we're not sure what all this neural activity is for?

Manzotti: On the contrary. We know very well that neural activity controls behavior, the nervous system having evolved to meet complex external circumstances with appropriate reactions. The question is, did it also evolve to orchestrate an internal mental theater for us David Chalmers's "movie in the head"? Or to "process information"? Stanislas Dehaene and Jean-Pierre Changeux, two leading neuroscientists, recently claimed that to explain consciousness we must show "how an external or internal piece of information goes beyond nonconscious processing and gains access to conscious processing, a transition characterized by the existence of a reportable subjective experience." There's barely a word here that refers to anything physical.

Parks: But is it really not possible to connect the notion of information with chemical exchanges occurring in the brain? Surely when we use a computer the information input is moved along toward the output through electrical signals. Can't this also be the case with the brain? Hasn't the philosopher Luciano Floridi claimed that "information is a physical phenomenon, subject to the laws of thermodynamics"?

Manzotti: Listen, when something physically exists and obeys the laws of thermodynamics, then you can find it, concretely. Electrons were predicted to exist and then found. Likewise the planet Neptune and a host of other things. But information, or data, is not a *thing*. It's an idea we stipulated because it served a certain purpose, but it doesn't exist *physically*, as an entity in its own right in the causal chain. Brutally, when we look inside a computer, or a brain, we don't *see* or even detect information. Or data. We see physical stuff: voltage levels in a computer, chemicals in the brain.

Parks: So what you're saying is that everything that goes on in a computer or in a brain could be fully and properly described without resorting to words like information or data?

Manzotti: Absolutely. Imagine you're describing a battery; you will have to refer to electricity. It is an indispensable part of the thing. But, when you describe what the brain or even a calculator does, everything can be exhaustively described in terms of causal processes, chemical releases, and voltage changes without ever using the word information.

Parks: But then what *is* information? How can Floridi make the claims he does? What part can information have in the consciousness debate?

Manzotti: Obviously there is the definition of the word in common use: "facts, data, communicated about something." The bus leaves at six. Yesterday it rained. The cash machine is out of order. That meaning has been around in English since the fifteenth century.

Parks: And?

Manzotti: Then there is the technical IT definition established by the mathematician Claude Shannon in 1949. Shannon was concerned about achieving accurate communication through technological devices and described information as an estimate of the probability that a given channel would successfully transmit words, images, or sound between a source and a receiver.

Parks: Sorry, what do you mean exactly by a channel? I'm lost.

Manzotti: A channel is the physical structure or circumstances that allow two separate events to be connected—the air pressure waves that occur when Romeo utters loving words to Juliet, the wire between a switch that you flip and a light that turns on, or everything that happens between your typing some letters on your phone and someone else reading them on theirs. Essentially, Shannon broke down any communication of data into its most basic constituents, namely a multitude of yes/no questions, that he called bits. Eight bits would make a byte. Information, in this new manifestation, is expressed as a number that tells us how many yes/no questions can be asked and answered through a given channel. A megabyte, for example, indicates capacity for around eight million such questions. Your smartphone requires a few million bits—yes/no questions—to put together, point by point, a photo on the screen. But there is no internal semantic content, no data or image *inside the device,* no point along the causal chain where you can put your finger and say, "Aha, information!"

Parks: Could we say that there is no more information in a cell phone than there is information in the air between my voice speaking and your ear listening? Or between a radio transmitter and a radio receiver?

Manzotti: You could, indeed. Information here is simply the capacity of any channel to effect a causal coupling between two events, speaking and hearing, typing letters and reading them. It is not a thing *between* those events. If there is no one on the receiving end to hear the voice or read the letters, then quite simply there is no information, because we don't have our two events.

Parks: So what do neuroscientists mean when they talk of information *processing* in relation to the brain? For example, a mouse's brain when the animal smells a piece of cheese.

Manzotti: Honestly, it is a bit like when we say that the sun goes down. Of course, we know it doesn't literally go down, but it is a nice expression and it saves a lot of explanation. The problem with the concept of "information" comes when we start to take it literally, as Floridi does. We start to imagine there really is mental, non-physical stuff called information. A subtle dualism creeps in, as if the brain contained organic material on the one hand and this mysterious, immaterial "information" on the other. In fact, Floridi speaks of moving from a materialist vision "in which physical objects and processes play a key role, to an informational one," as if there were some sphere of existence that is not physical.

However, in its precise scientific usage, and certainly most neuroscientists would see it this way, "information processing" simply means that a physical system—a computer, or the human body, the brain—allows given events to pass along their causal influence to further events. When your mouse recognizes the smell of cheese and moves toward it, the cheese becomes the cause of effects in the olfactory bulb, which themselves cause effects in the lateral entorhinal cortex, which themselves cause effects in the hippocampus, and so on. But there is no immaterial "message" being passed along, no code, or coded representation of "cheese," existing separately from these organic changes, which are very many and very, very complex. The notion of information and information processing is then built on top of all that causation. It is a kind of shorthand for describing a causal chain so complex as to be beyond any visualization or easy explanation.

Parks: But does the chain end anywhere? Is there a point where we could say, this is where everything arrives, where conscious experience happens?

Manzotti: Alas, no. Rather than ending, the causal chain branches and every branch is a constant back-and-forth with as much feedback as input. In this sense, the brain is completely different from IT devices which are always channels leading somewhere, usually to a person who reads the message that arrives—the second of the two events we talked about.

Parks: Okay, let me try to sum up so far. The neuroscientists, for the most part internalists, continue to fill us in on the brain's exceedingly complex chemical and electronic activity. Meantime, the extended computer metaphor that they almost always employ conveys the impression that what is going on is not just organic, but "mental," that the brain is producing consciousness, storing memories, decoding representations, processing data. So there is a general feeling of promise and expectation, but actually we get no nearer to an explanation of consciousness itself, since we are simply describing, with ever greater precision, what neurons organically do.

Manzotti: I'd agree with that. And perhaps add that maybe people are not unhappy with the situation: we get regular, often melodramatic updates on how marvelously complex we are and how clever scientists have become, while consciousness remains blissfully mysterious. In short, we get to feel very special all around.

Parks: Let's stick to substance. Aware of this situation, some internalists have made other suggestions. David Chalmers, if I'm not mistaken, has suggested a sort of second and secret life of information—hopefully you can explain. Giulio Tononi has developed an elaborate theory of "integrated information" and "emergence."

Manzotti: Both Chalmers and Tononi seem to see information processing as a sort of intermediate step toward the conscious mind. I'm not sure this is very enlightening, because if it is hard to imagine how consciousness might "emerge" from neurons, it is even harder to conceive how it might "emerge" from information, which, as we said, is not a physical thing, not "a thing" at all in fact. To put it another way, information can hardly form the basis for a natural phenomenon like conscious experience, which—and we must always remember this—is a *thing*, a physical phenomenon that we all experience at every waking moment.

Parks: Let's take the positions one by one. What is this dual aspect of information that Chalmers proposes?

Manzotti: Chalmers agrees that information, as Shannon construes it, lacks any phenomenal character (colors, smells, feelings), or indeed intrinsic meaning, in that a string of zeros and ones in a computer might mean anything. Yet he believes that the brain is basically a computational device crammed with information. So how do all those zeros and ones, or some neuronal version of the same, become colors, sounds, pains, and pleasures? His solution is that information has a *dual aspect*—the *functional* aspect (the zeros and ones that govern our behavior) and the *phenomenal* aspect that constitutes conscious experience (colors, sound, itches, whatever). He does not explain why or how this should be and admits himself that his position is basically dualist: information has two sides, one that science can deal with—neurons controlling behavior—and another that is, simply, consciousness.

Parks: At which point we're back with Descartes deciding what belongs to science and what doesn't.

Manzotti: Pretty much. Tononi also distinguishes between two kinds of information. Standard information, of the IT variety, and "integrated information," which we find in the brain and which, like Chalmers's second, "phenomenal" aspect of information, gives rise to consciousness.

Parks: But what is "integrated information"?

Manzotti: It's a model for quantifying how much any system brings together, or integrates within itself, the causal influences of the external world. For instance, in a starfish, none of the separate arms knows what the others are up to or what is happening to them. There is no integration at a neural level. Or consider an image on your computer screen: each pixel is quite independent from the pixels around it; you can change one without altering the others. Human beings are very different. Change one neuron and changes will occur in hundreds, if not thousands, of others. Read about Joyce's Stephen Dedalus in *Ulysses*, or Proust's Marcel in *À la recherche du temps perdu*, and you'll see that everything that happens to them is immediately mixed up with everything else. Everything connects. A human being is the ultimate causal Gordian knot. You can't disentangle it. So Tononi's integrated information is a formula that expresses quantitatively the extent of such integration in different creatures and systems.

Parks: Does that mean it can calculate how those creatures or systems react to a given stimulus?

Manzotti: Perhaps potentially yes, but at the moment no. Tononi's formula is so complex to compute that even if you posit an unrealistically simple nervous system, it is still beyond the capacity of the most powerful computers to handle. Aside from that, the formula does not explain how or why this super integration might transform itself into the things we experience, for example, the color red.

Parks: It seems whichever way internalism turns, however exhilarating its interim discoveries, when it comes to consciousness it reaches an impasse. We have the impression—or simply we're used to believing—that consciousness is in our heads, that memories are stored in our brains, that there is a world outside and a representation of the world inside, and so on. Yet nothing we have found in the brain warrants this. In our next dialogue, then, I propose that we break out of our skulls and see if there is any other approach to this question that offers more promise.

Manzotti: Very good, but this time I'm going to have the last word! Internalism, like dualism, is, if you'll allow me the joke, a monster with many heads. We're going to have to come back to it again and again, to look at dreams, visualizations, hallucinations, and all kinds of other exciting creatures. And some of them will be harder to tackle than the basic premise in itself.

4. THE ICE CREAM PROBLEM

The average human brain weighs in at something under three pounds and has a volume of 1,250 cubic centimeters (76 cubic inches). Despite the complexity of its architecture and the daunting interconnectedness of its 85 billion neurons, the goings-on in this small space have now been pretty well documented. We know what faculties are impaired when each part of the brain is injured, which neural activity, more or less, correlates with which behavior. Yet, as we discussed in our earlier dialogues, all these impressive results have not brought us any closer to accounting for consciousness or even establishing where exactly it "happens."

How have scientists and philosophers dealt with this impasse? Some, like the philosopher Galen Strawson, have suggested that since it is self-evident that consciousness is real, and equally evident that neuroscience hasn't accounted for it, there must be crucial things we don't know about the physical world. Others, like the neuroscientist and psychiatrist Giulio Tononi, suppose consciousness must "emerge" from the highly integrated neural processes taking place in the brain, yet, as we noted in our last conversation, they don't seem to have any conclusive empirical evidence. Others still, like the philosopher and cognitive scientist Daniel Dennett, deny that consciousness exists at all and insist that integrated selfhood is a delusion. While

*many of Dennett's points make good sense, his claim that the
issue of consciousness is mostly a matter of conceptual confu-
sion has not brought us any closer to understanding the nature
of conscious experience.*

*There are, however, a number of scientists and thinkers who've
taken a different line. Not convinced that the consciousness
problem will be solved by studying the brain alone, they have
begun to look outside the head.*

—Tim Parks

Tim Parks: Riccardo, how can one conceive of consciousness if
it is not a brain-generated representation of the world outside?

Riccardo Manzotti: Let me offer a premise. I believe we are up
against two equally strong, equally commonsensical, but incom-
patible intuitions. We feel that we, our selves, are located where
our bodies are, and very likely *inside* our bodies. On the other
hand, we don't feel we are made of the kind of stuff we see when
we look inside a human body. Our conscious experience is of
quite a different nature from these cells, membranes, muscles,
fat, and bodily fluids.

Parks: Could that be why we're so fascinated by movies, paintings, even anatomy drawings that show the body being cut up or
taken apart?

Manzotti: I've often thought so, yes. The real horror is not what's
there, the gruesome mess, but what's *missing*! Rummaging through
the body's innards, we don't see anything that resembles a self.

Parks: Traditionally the way around this has been to suppose that the mind, or even soul, is indeed in there, but invisible, insubstantial—made of non-physical stuff. Surely modern science couldn't be accused of taking this position.

Manzotti: You would think not, yet recent neuroscience is perilously close to it. Consciousness, the neuroscientists claim, is inside the brain but eludes our observation. They see neural activity that correlates with consciousness, but acknowledge that this is not consciousness itself and don't even try literally to observe consciousness itself. Since everything that is physical is *detectable*, this is akin to claiming that consciousness is not physical. And so we go back to Descartes and to dualism.

Parks: Let's move on, then, to the obvious alternative: that we've been looking in the wrong place. Is the idea that the brain might not be the seat of consciousness a new one?

Manzotti: Not at all. If you like you could go as far back as Aristotle, who claims that the soul is, if only briefly, *identical* with the objects of our experience: the soul is "in a way all existing things; and knowledge is in a way what is knowable, and sensation is in a way what is sensible." It's worth remembering that for Aristotle the soul is material rather than spiritual, and thus akin to what we mean today by consciousness. So what he's saying is that, at any given moment, our consciousness is identical with the form of what it is conscious of. When you see an apple, your consciousness and the apple are made of the same stuff, which he calls the form of the apple. If only, as he says, briefly.

Coming nearer to our own time, the behaviorists, B. F. Skinner in particular, argued against consciousness as something internal and against memories as things that are stored in the head. Like Dennett later, they played down the importance of phenomeno- logical consciousness and concentrated on manifestations of self- hood in behavior. Essentially, they were reacting against notions such as "introspection," "inner mental life," and other remnants of German idealism, not to mention various unscientific forms of spiritualism. Yet by focusing exclusively on observable behavior, they threw the baby out with the bathwater, ignoring conscious- ness altogether.

Parks: But where would consciousness be, exactly, if not in the head? What would constitute it?

Manzotti: In the second half of the twentieth century a number of thinkers, notably the psychologist James J. Gibson, began to focus on the interplay between body and environment. Rather than a representation of the world in the brain, a "movie in the head" as it were, perception was conceived as a skillful activity, or interaction with what is around us.

Parks: I do therefore I am!

Manzotti: Exactly. These thinkers are not denying or ignoring consciousness, as the behaviorists seem to be doing. Rather they talk a lot about our initial experience of the world, how as babies we discover our environment through touch and exploration, how all experience amounts to an engagement with what the world offers us.

Parks: Tell us who these people are. Talk us through it.

Manzotti: There are many versions of the same idea and plenty of fancy names: ecological perception, embodied cognition, externalism, enactivism, and the extended mind. Essentially, though, despite their many nuances, all of them are making the same claim: that what the body *does* constitutes, causes, or is the basis of the mind. Gibson, for example, worked with Air Force pilots and suggested that consciousness might be identical with the interplay between body and airplane or even with the interplay between plane and airstrip. Crucially, he introduced the notion of "affordance." This is the idea that every external object offers or "affords" the body certain opportunities for active engagement. Crudely, you could think of how a pair of scissors offers itself to the thumb and fingers. Notice that this "affordance" depends on both body and object. For an animal that didn't have the *opponens pollicis*—opposing thumb and fingers—the scissors would offer quite different possibilities and thus different affordances. Doorknobs, steps, bicycles, and keyboards all offer obvious cases of coupling between the body and external objects. Gibson's point was that there was no need for these "affordances," this matching of body and environment, to be represented in the brain, since they were already present externally in the meeting of body and object.

Parks: I'm sorry, but I can't help noticing that all the examples you've given are man-made objects. Things made on purpose to fit our bodies.

Manzotti: I was trying to get the idea across as clearly as possible, but you're right, there's a problem here. Man-made objects are designed to have specific affordances that pull you toward them. In principle, you can see the affordances that, say, pebbles or bananas might offer a human being, or the ground beneath our feet for that matter. But such affordances are less clear in nature. Then of course there are many things—distant mountains, clouds in the sky—with which we have no *interaction* at all, yet we experience them just the same. Still, the biggest problem for Gibson's approach is that the very idea of action is "mental."

Parks: I don't follow you.

Manzotti: Well, we can all see the persuasive power of the baby coming to consciousness through interaction with the external world. But we describe the child's explorations as "actions" because we assume that he or she is a *subject*. What I mean is, the notion of action requires that first we have a subject, a conscious being. You wouldn't say that a washing machine *acts*, but a person washing clothes does. A dog acts, or even a mouse, but not a glacier, or a wave moving across the sea.

Parks: So if the notion of action *depends* on consciousness, you can hardly use it to account for consciousness.

Manzotti: Right. However integrated we may be with our environment—and I believe we are—actions are not special things that can be used as building blocks to manufacture consciousness. They are events that happen because subjects do them to pursue their goals.

Parks: So this approach was short-lived?

Manzotti: The initial intuition that consciousness *requires* the external world we experience every day, that the mind isn't simply locked in the skull as the internalists would have it, is a powerful and persuasive one; hence, after Gibson, there were lots of people eager to find ways around the problems this approach faced. In a seminal paper on behavioral and brain sciences published in 2001, J. Kevin O'Regan and Alva Noë offered a new version—they called it *enactivism*—that made a big impression on me at the time.

Parks: Why was that?

Manzotti: First, they explicitly abandoned the idea that perception involves representations of the outside world created inside the brain by our neurons. That took courage. Instead, they proposed that seeing is a form of physical action. For example, they drew attention to the fact that our eyes are always in movement when we are looking, even though the object we are looking at appears to be still. I was impressed, in particular, by their claim that the skin is not some kind of magic threshold or insuperable barrier; on the contrary, there is a seamless, physical, causal continuity between what's going on in the brain and what is happening in the world. When you see, photons bounce off the objects seen and pass through the air to meet the retina. When you hear, sound waves agitate the eardrum. And when you hold a tool in your hand, they pointed out, it actually becomes an extension of your experience. The Cartesian spell that kept subject and object apart, something the neuroscientists had reconstituted as the separation between brain and world, seemed on the point of dissolving.

Parks: It's good to hear you enthusiastic for once. But I'm afraid I don't get it. I can see that consciousness might be able to encompass certain prostheses. In the end, even a pair of glasses becomes part of our system of perception; a pen or keyboard often seems an extension of oneself and I can imagine a seamless causal chain moving from my neurons to the pixels on my screen. I also take the point that our eyes are constantly mobile as they engage with the world. But I don't see how that actually accounts for consciousness, for our experience of the world, or even of our bodies. I must say I've read Noë and really enjoyed his lively approach, but this was my problem throughout.

Manzotti: Well, you're right to be skeptical. O'Regan and Noë ditched the term affordance, which carried a hint that the object was designed specifically to integrate with human action, and spoke instead of objects offering "sensorimotor contingency." But essentially it meant the same thing and didn't solve the fundamental problem: if consciousness is constituted by actions, then for every experience there *has to be* a corresponding action. This sounds fine when we're talking about things like surfing, skiing, carving the Sunday roast, or even kissing, but there are so many cases where it doesn't work.

For example, I lick a strawberry ice cream and a chocolate ice cream. The action, licking, is the same, sensorimotor contingency, that is the affordance that the two ice creams present to the mouth, is the same, but the taste quite different. There simply isn't a different *action*, a different engagement of the body with the environment, to match every different experience we have.

Parks: I presume we're distinguishing, then, between the action of licking and the action, if you want to call it that, of the taste buds as they meet the different flavors?

Manzotti: From a mechanical perspective, different taste buds do not perform different actions. They are triggered by different molecules and are, of course, connected with different neural groups. But saying that a bitter taste is the outcome of bitter taste buds is tantamount to saying, as Johannes Peter Müller said in the nineteenth century, that the quality of our experience is provided by the peripheral nerves, an idea now entirely discredited, first and foremost by neuroscientists. The problem for the enactivist theory is that the tongue performs the same action no matter what ice cream is melting on it. And then that the properties of the experience—whether a chocolate taste or a strawberry taste—are different from the properties of the physical action, licking.

Parks: All the same, action does shape and enrich our experience, doesn't it? And we know that every action, every perception, subtly alters neural connections and pathways in our brain, such that we recognize at once when we have already done something before, touched a particular surface, tasted a particular ice cream.

Manzotti: Who could disagree? The problem is that what philosophers and scientists propose as the basis of consciousness—be it neurons or action—invariably turns out to be just one of many elements involved in delivering or tuning consciousness. Not the thing itself. Neurons obviously have a part; damage them and your consciousness will be changed. Actions have a part; as you say they shape and enrich consciousness. Culture and society also have a part; they mold and direct our consciousness. But none of these things is consciousness itself.

Parks: How did the enactivists handle dreams?

Manzotti: Well, here we have the obvious problem of experience without action. When we dream, we are for the most part lying motionless. The eyes may be moving beneath the eyelids, but they are certainly not interacting with affordances in the immediate environment. In any event we now know that while most dreams take place during REM sleep, we can in fact dream in all phases of sleep. And of course, dream experiences are hardly limited to the visual. In order to explain this special case, then—experience without action—some enactivists have begun to distinguish between *potential* actions, or *dispositions to act* in response to the world's affordances or sensorimotor contingencies and real actions or *enacted* actions.

Parks: So dreams become just one category of potential actions?

Manzotti: That's it. When you are experiencing but not acting you are *potentially acting*. The problem is that in suggesting this solution they have created something separate from the physical world, a shadowy layer of possible, hypothetical actions waiting to be brought to life.

Parks: It sounds like we're heading back to a separation between mental and physical, where the mental is everything we're unable to account for.

Manzotti: To make matters worse, Noë speaks of our "possessing sensory-motor knowledge" rather than just interacting with the environment. This means that consciousness has now been relocated to a sort of abstract level—knowledge—which looks suspiciously like a *mental* repertoire of stored representations. It hardly matters whether these are representations of actions or sensorimotor contingencies, rather than objects, they are still representations, namely something different from physical reality. This was precisely the kind of theory enactivism had set out to debunk.

Parks: It really does seem impossible to think about consciousness without falling back at some point into this Cartesian view, the real world out there and a representation of it in the head.

Manzotti: You can see why everyone is willing to give so much credit to the neuroscientists, or just scientists in general, hoping they will come up with something that will solve the dilemma, some as yet unknown aspect of the material world that will explain why consciousness is indeed in the head, but has nevertheless managed to remain invisible up to now.

Parks: I can see you don't believe that's going to happen. But at least it's now clear that behind the problems encountered by all our thinkers, of whatever persuasion, is the vexed relationship between subject and object, the conundrum of how the world can be both out there *and* in our heads, apparently, at the same time. You feel that the enactivists' decision to look for an answer outside the head was right, but that the focus on action was wrong. Let me propose then that in our next conversation we tackle that issue head-on: subject and object, inside and outside. Let's consider every possible relation between the two. I also think, Riccardo, that it's time you showed your hand and told us what you think. It is too easy breaking down other people's ideas; you should give us something positive to chew over.

Manzotti: It will be a pleasure. But let me invoke Sherlock Holmes: "How often have I said to you that when you have eliminated the impossible, whatever remains, however improbable, must be the truth?"

Parks: Excellent. So in our next conversation we dispatch the last impossibilities and expect to be surprised.

5. AM I THE APPLE?

How is it that we experience the world? How is it possible that the environment we live in, the objects we use and see, touch and taste, hear and smell, are both patently out there and simultaneously, it seems, in our heads? After four long conversations, considering the positions of philosophers and neuroscientists, those who assume that experience is an amalgam of neuron-generated representations in the brain and those who have looked for it in our interaction with the environment, Riccardo Manzotti and I are no nearer to establishing what consciousness is or where it resides. Today, then, we have set ourselves a simple task: to review all the ways philosophers have supposed a subject might relate to and become conscious of an object, setting aside once and for all those hypotheses that have clearly failed and asking, "Is there one approach which has not yet been given due attention?" Riccardo believes there is.

—Tim Parks

Tim Parks: Riccardo, to talk us through this I know you want to propose something you call "the metaphysical switchboard." Can you explain?

Riccardo Manzotti: Well, at the beginning of any discussion of consciousness there are some fundamental premises to be established that will constrain everything that follows. The metaphysical switchboard will help us get a grasp of those premises and the various directions they lead in.

Parks: I'm all ears.

Manzotti: So, imagine an old-fashioned switchboard with just two toggle switches. Each time you flick one of those switches you open a path that sends the debate in a different direction. Let's say the first switch determines whether or not subject and object are to be considered *separate*, the second whether or not the subject is to be supposed *physical.*

Parks: This should be fairly straightforward: two toggle switches means just four available pathways. Let's start by turning our first switch in favor of a separation of subject and object, since I'm sure most of us think of our minds as separate from the objects they perceive.

Manzotti: Fine. With the first switch on *separate*, we now have to set the second. If we set it for a subject that is *non-physical* then we get Descartes's immaterial, or spiritual, subject. That is, in the Cartesian model, we have immaterial souls, or just selves if you like, separate from the physical objects we experience and even, ultimately, from our bodies. It's a solution that has occupied a huge space in modern history and that many religions endorse, but scientifically it's a non-starter, since it's based on the notion that the subject *cannot* be an object of scientific enquiry. So we can be forgiven, I think, for paying it no further attention.

Parks: No souls flying up to heaven, freed from their material prison.

Manzotti: Alas, no. All the same, assuming we now flick that second switch to *physical*, things hardly get much easier. This is the territory, I should say, of modern science from Galileo right up to today's neuroscientists. They left the first switch set to keep subject and object separate, but placed the subject in a predetermined place in the physical world, namely the brain, and hence made consciousness a neural process that is inside the head and separate from the physical world it perceives. Unfortunately, since all available empirical findings have shown that the properties of neurons are nothing like the properties of our minds or our experience, this shows every sign of being another dead end.

Parks: We discussed the brain-based solution in our first dialogues. But did setting our second switch to *physical* mean we had to place the subject in the brain? Couldn't it have been the whole body?

Manzotti: Some philosophers and scientists—Francisco Varela, Maurice Merleau-Ponty, and Rodney Brooks, for example—take the whole body and what the body does to be the basis for the mind, but, at least as far as regards consciousness, they've had no more success than those focusing on the brain. You can look at muscles, blood cells, and nerve endings for as long as you like without finding anything that remotely resembles consciousness.

Parks: Time to go back to the first switch then and set it for a subject and object that are *not separate.* Is this, perhaps, what the enactivists we discussed in our last conversation are doing? Is this what they mean when they suggest that consciousness is not contained inside the head but constituted by our interaction with the world?

Manzotti: The enactivists toy with the first switch, without actually turning it all the way to *not separate.* They see that consciousness can't be reduced to a property of the goings-on in the brain, so they start to look outside. But instead of considering the external object as such, they look at our dealings with the object, our handling the object, our manipulating the object, believing that consciousness is a product of the actions we perform. At the end of the day, though, the object remains doggedly separate from the subject who experiences it. And unfortunately, as we said last time, actions, whether they be eye movements, or touch, or chewing, are no better than neural firings when it comes to accounting for experience. How can my actions explain why the sky is blue or sugar sweet?

Parks: Okay, let's stop playing with that switch and set it determinedly on subject and object *not separate.* As for the second switch, let's again start with a subject that is *not physical,* since I suspect you are going to give that position short shrift.

Manzotti: Yes. This is the territory of Bishop Berkeley and Gottfried Leibniz in the late seventeenth and early eighteenth centuries. Crudely speaking, they proposed that subject and object become identical, the same thing, but both in a completely non-physical world.

Parks: What exactly do we mean by identical? How can different things be the same thing?

Manzotti: Well, identical the same way that Bruce Wayne and Batman are identical. Bruce Wayne *is* Batman. For a true idealist—that's what this approach is called—the object of perception is nothing but a modification of a part of the subject who experiences mental representations or ideas, which are not physical. The world is all idea. Actually, you could say that this position has recently been revived by techno-enthusiasts like Elon Musk who wonder whether we live in a giant computer simulation concocting the illusory objects we erroneously imagine reality is made of. These are intriguing ideas, but again they are scientific non-starters since they depend on a non-substance which no one can track down or verify. You can't even begin to prove them wrong.

Parks: In our last conversation you reminded us of Sherlock telling Watson, "When you have eliminated the impossible, whatever remains, *however improbable*, must be the truth." And I guess we've reached that point: three of our four pathways have been dismissed, so let's turn to the last: first switch on *not separate*, second on *physical*; that is, subject and object are identical—the same thing—*and* physical. Improbable indeed! Crazy, most people would say. The only philosopher I can think of who has given a nod in this direction is Aristotle, when he says that the mind "is in a way all existing things." But I suspect most of our readers will take that to be esoteric rambling from thousands of years ago.

Manzotti: Let's forget the voices from the past, wipe the slate clean, and take a really ordinary situation: a person looking at some everyday object. Let's say, an apple. That's the example I always use. So, we have a body—including a brain, of course!—and an apple. Nothing esoteric or improbable there. The body, let's say your body, on the left, the apple on the right, thin air in the middle. What could be simpler? But there's still one piece of the jigsaw we haven't placed yet—your *experience*, or simply *the experience* of the apple. *Where* is that? And *what* is that? We can't account for it. So let's consider the chief suspects, one by one. Is it the activity of your neurons?

Parks: We've already decided against that. There is nothing applish in the gooey brain, and anyway, we don't experience neurons, we experience the apple.

Manzotti: So is your experience some movement you're making in relation to the apple? Some action? Is it your movements that conjure a sort of applishness?

Parks: Again, we've already dismissed this, haven't we? Movements, actions, don't seem to have anything in themselves that we can identify with the apple we see. I can't imagine a robot repeating exactly my movements would have the same experiences I have. I think we can let this go.

Manzotti: Okay, so could our experience of the apple be an amalgam of everything going on between subject and object? Neural processes, retina, optic nerve, molecules of apple, atoms in the molecules, electrons in the atoms, everything?

Parks: You're trying to treat me like a performing dog, Riccardo. You want me to say no, obviously. Instead I'm tempted to say yes. Why can't consciousness be the whole process?

Manzotti: Because just as we don't experience neurons, so we don't experience retinas or photons either. Obviously they're necessary to the experience of seeing an apple, all the elements of the process are necessary, but they're not *it*, are they?

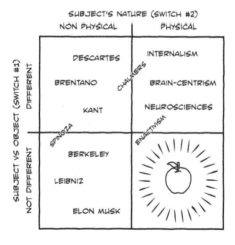

Riccardo Manzotti's metaphysical switchboard.

Parks: Not convinced. Just because we don't experience the constituent parts of a system, it doesn't mean that they aren't together what we are experiencing.

Manzotti: The constituent parts are causally necessary the way a pot is necessary to boil pasta, but the pot is not the cooked spaghetti. Nor is the heat source, nor is the water. In the case of the apple, if you list the properties of the process as a whole, they just don't match the properties of the experience. Photons are not applish. The rhodopsin your retina secretes is not applish. And so on.

Parks: I need to think more about this. All I'm going to concede for the moment is that when I experience the apple, I'm not aware of experiencing the other various elements of the process.

Manzotti: Okay. By all means, think it over. But meantime what if someone said that the experience was *the apple itself*? After all, the apple is definitely the most applish thing around. And the only thing that has the properties of an experience of an apple. It's round, it's red, it's shiny. So why can't the subject, consciousness (*but not the body*, notice), be identical with the apple, out there where the apple is? Consciousness is not *about* the apple, consciousness is the apple.

Parks: Of course, I've heard you come up with this argument before, Riccardo. So I'm not going to pretend to be amazed. But it's still extremely difficult for me, and for most of our readers, it will seem quite mad. Essentially, you're proposing that my experience of the apple, or any "external object" is *outside my body*—out there where the object is.

Manzotti: Exactly. But did you ever really imagine the apple was *in your brain*? That it somehow got smuggled in among your neurons? Of course you didn't. You always thought the apple was out there. And so it is. And with it your experience.

Parks: Again—because I still find this a huge conceptual struggle—you're saying that my experience is literally the object. So my consciousness, the me, the subject, or at least the *part of me* that is the experience of the apple, is *identical* with it.

Manzotti: Right. Experience is physical—what else could it be?—only it is not the physical object that it is usually assumed to be, our neurons, but another physical object, the apple.

Parks: But in that case, what would be the relation between my body, my brain, and this apple experience that is halfway across the room? And what is the apple when it's not identical to my experience? Does it disappear with Bishop Berkeley's trees in the woods when no one's looking at them?

Manzotti: One objection at a time! But first, please note that there is nothing here that contradicts the findings of neuroscience or, indeed, physics. All that we know about what goes on in the brain, all the correlations between neural activity and specific kinds of perception, all the physics of photons and sound waves, all the chemistry of retina and taste buds, all the mechanics of the ear and the nose, remain absolutely in place. Everything is physical, verifiable. We just have this one, admittedly enormous, conceptual shift: instead of supposing that the senses receive "input" and somehow create a second, inner mental world reflecting the outer world, we say that your experience is in the outer world; it is not separate from the physical object you perceive, *it is the object.*

Parks: I can see it all seems extremely simple to you, and that you've somehow convinced yourself it's true. But I assure you that for most people this idea will seem bizarre and almost mystical. Please answer my previous question: What is the relation between my body and the distant object, which is also, you say, my experience?

Manzotti: Francis Bacon remarked that "opportunity makes a thief." Likewise, we could say that your body offers the opportunity or physical conditions—eyes, optic nerve, neurons and so on—that allow the world to take place as the object we experience.

Parks: Opportunity makes the object.

Manzotti: If you like. Of course, the apple conjured up by this opportunity that is your body, the apple you perceive—is not an absolute apple, it's not the apple that Galileo supposed was altogether measurable and fixed, nor Kant's *noumenon*, the apple-in-itself that can never be known; it's not an apple in an X-ray machine, nor a slice of apple under an atomic microscope. The apple that you experience is simply a selection, or subset, of the many other things going on out there in the world; it is the selection that your body—your brain plus your sense organs—allow for.

Park: A relative apple.

Manzotti: Right. It's relative to your body. Though of course, since most humans have similar perceptive equipment, we will tend to agree on shape and color up to a point, depending on our eyesight, our position, and so on. Other animals or other devices will allow for other selections and hence other object experiences, which are equally relative and equally real. Your apple is not a snail's apple. Or a bat's. Your apple is made of those *and only those* physical features that cause effects thanks to your sensory organs, your particular body. Which is not to say that other properties do not exist. They do. It's just they're not part of the object that you experience; to wit, one side of a shiny red round apple.

Parks: So I am the apple.

Manzotti: Of course that sounds absurd, because you identify your conscious self, the subject, the I, with your body, and your body is clearly not the apple. But what if I were to say that the very idea of consciousness was invented to explain how you could experience an apple when there is no apple in your head? So we have to have this consciousness apple. However, if experience and apple are one and the same, there is no longer any need to talk of a consciousness separate from it. The apple is more than enough.

Parks: You're really going to say I am the apple...

Manzotti: You are a whole range of experiences, hundreds and hundreds of things going on simultaneously, of which the apple is one. That's why when you close your eyes, the apple disappears. Eyes closed, your body no longer offers the conditions for the apple to have certain effects. Consequently, the apple with its shape and colors is no longer part of your experience.

Parks: But I know it's still there! I could reach out and touch the apple, eyes closed.

Manzotti: You could, indeed! But it will not be the same apple as your visual apple. It will be a smooth solid rotundity. A blind man's apple. And that apple too will be in the external world and not inside your hand. This "touch apple," if I can call it that, originates in the same *external conditions*, but a different set of physical features is now selected. Again the object is *relative* to the body, or the parts of the body, involved in the experience. It's like the difference between feeling for something inside a drawer as compared to looking inside a drawer. Different experiences, different objects. None of them in your head, but out there, where you experience them.

Parks: So, if I'm not mistaken, what you're claiming is that for every experience *there must be an external object which is identical with it.* And frankly that is not going to be an easy sell. People will laugh you out of town. What about memories, they will say? My memory of the apple. What about dreams? What about hallucinations? And thoughts, words, cogitation, pain, heat and cold. Where is the external object that corresponds to these aspects of experience? Your idea is dead in the water.

Manzotti: By all means, bring the objections on. I'm ready for them. All I ask is a little time and space. You can't turn round an oil tanker on a dime. We'll tackle these challenges, which are substantial and serious, in our next talk. But let me just say in closing that this is not simply philosophical speculation, but a concrete empirical hypothesis. It's a risky hypothesis, I know, but didn't Karl Popper define scientific hypotheses as inevitably risky, daring proposals open to being proved or disproved?

Parks: Which you believe this is?

Manzotti Eminently.

Parks: Okay. Next time, we'll give you space to say your dime's worth, but let me warn you that I'll be wanting to know where exactly is the external object that corresponds to each of my experiences.

6. THE MIND IN THE WHIRLWIND

In our last conversation about consciousness, Riccardo Manzotti and I arrived at a crux. Having found both brain- and action-based explanations of conscious experience unconvincing, Riccardo set out a radical alternative: our experience of the world (light, color, sound, smell, touch) is not a "movie in the head" provided by our neurons, nor the interaction between our bodies and our environment, but nothing other than the object itself. When I see an apple in front of me, I am the apple. Every perception is nothing more, nothing less, than the object perceived, hence every experience requires an external object to which it corresponds.

To those of us used to supposing that our experience is locked inside our heads and that our minds begin and end with our bodies, this externalist approach initially seems completely unacceptable, even laughable. How can my consciousness be both physical and outside my body? How can a subject, I, be identical with a thing? And since my experience changes, but the object clearly doesn't, how can the two be the same?

Yet, casting around, one finds corroboration in surprising places. "I am what is around me," wrote Wallace Stevens in

his poem "Theory." "You are the music while the music lasts,"
reflected T.S. Eliot in "The Dry Salvages." "What are called out-
side and inside are one and the same," wrote Samuel Beckett
to Georges Duthuit. Virginia Woolf's Mrs. Dalloway is a con-
stant elaboration of this intuition: "She waved her hand, going
up Shaftesbury Avenue. She was all that." Nor are artists alone
in arriving at these reflections: "Nothing can represent a thing
but that thing itself," thought the philosopher Edwin Holt in
1914. Two thousand years before him, Aristotle claimed that
"actual knowledge is identical with its object."

Such fragments hardly amount to proof, or even a hypothesis.
But they do suggest that to consider experience as one with
the object experienced is not perhaps so counterintuitive after
all. What we want to do today is flesh out what Riccardo calls
the "Mind-Object Identity Theory" and confront some of these
immediate objections.

—Tim Parks

Tim Parks: Riccardo, at the beginning of these dialogues it seemed
that the principal focus of our enquiry was consciousness, our
moment-by-moment experience of the world. But in seeking to
explain that you have made a problem, even a mystery, of something we assumed was fairly simple: the objects we perceive, the
things we experience. You tell me that when I see an apple, my
visual experience of it *is* the apple. Hence the apple is me, or at least
it is that part of my overall experience that is the apple. My body,
with its sensory capacities, creates an opportunity for what is out
there to be the way I see it, at which point *the being is the seeing.*

At the same time, a dog looking at this very apple, or a bird, or just another person on the other side, sees something different. Yet each of them, you claim, when experiencing the apple *is* the apple. How can that be? And of course, if I move around the apple, or see it in a different light, my experience changes; so if I am identical with the object, then presumably the object also changes with my experience. There is no stable apple. Have I understood? And if I have, what sense am I to make of this? Doesn't science, and common sense, tell us that an apple is an apple is an apple? Can't it be measured, photographed, analyzed in a hundred possible ways? Its physical properties don't change, do they? It is what it is.

Riccardo Manzotti: Your objection is that the world appears to change while actually you feel sure it remains the same. Right?

Parks: My experience of the world is different from moment to moment depending on where I am, where I look, how the light is, etc., yet the world seems to remain reliably there and unchanged. So how can my changing experience *be* the stable object I see?

Manzotti: You realize, of course, that you're simply restating the traditional appearance/reality conundrum that has plagued human understanding since Plato. The world *appears* to be one thing, but I *know* it's another. Right? And we all know where that debate leads: appearances get relegated to an inner mental domain—these days the brain—while outside the world stays real and largely unknowable. What I'm asking you to do is set aside the idea that there are appearances on the one hand and real objects on the other. There are only objects. Real physical objects. Your experience *is* those objects. There is no appearance/reality dichotomy.

Parks: Of course, it's exciting making grand statements that dismantle thousands of years of philosophy, but that doesn't answer the question: How is it, if my experience is identical with the object—say your famous apple—that I can see one apple and someone else a different one? If we were both "identical" with the apple, surely the experience would be the same and *we would be identical with each other,* which is clearly nonsense. We both look at the apple but that doesn't mean I'm you. God forbid!

Manzotti: Okay, let's go back in time and do some basic thinking. Until the seventeenth century scholars believed that when a body or object was in movement, its velocity was to be considered an absolute physical property. A cannonball was either moving or still. How could it be otherwise? A flying bird is moving and a mountain is still. It's a no-brainer. Then Galileo came along and showed them they were wrong. Velocity is relative. The mountain is still with respect to the surrounding landscape, but it is moving with respect to the moon or indeed with respect to the flying bird. Eventually it was established that any and every object has infinite velocities, each relative to some further object. Velocity is always relative velocity.

Parks: This seems elementary.

Manzotti: Well, velocity is *physical,* is it not? It is not notional, it is not just "in the head," it is not *subjective.*

Parks: And so?

Manzotti: My point is that despite being physical, it can't exist *by itself*. It requires a relation with another physical system. You can't say, or at least not in scientific terms, that an object has this or that *absolute* speed; you need to say it has this speed *relative* to that object. So let me put it to you: What if all physical properties were like velocity? Not absolute, but relative to other objects. What I'm asking you to contemplate is the notion of *relative existence.*

Parks: So not only is the apple neither still nor moving except in relation to other things, now it doesn't even exist except *in relation.* You're going to have to work hard to convince me of this. In relation to what?

Manzotti: The temptation is to say, in relation to *you* or *me*, but in that case we'd be resurrecting an immaterial subject, an insubstantial self, and getting ourselves into all kinds of trouble. The apple is relative to another physical object or objects, in the same way the velocity of the mountain is relative to the flying bird. So, what object is the apple relative to? Many. If it's a case of experience, a body. A human body. An animal. An insect. But an apple also exists in relation to the surface it lies on, a table maybe, or the tree it hangs from. Bodies are not metaphysically special. They are just objects, though very complex objects, in the sense that, through their sensory capacities, they bring into existence the world we are identical with: the sights, sounds, smells, and so on that are our experience, are us.

Parks: *Bring into existence!* Surely an object exists regardless of the other objects or bodies around it.

Manzotti: A key is a key only relative to its lock. Relative to anything else it is just a piece of metal. A face is a face only when in relation to a healthy fusiform gyrus, that part of the brain we know is necessary for face recognition. The same physical stuff constitutes different objects, depending on the other objects it is in relation with. Just as it has different velocities relative to different objects. *Existence of this or that object is physical yet relative!*

Parks: Riccardo, the idea I'm getting is this: that you are working crazily hard to eliminate anything intermediate between the body and reality. You don't want appearance, you don't want representations, you don't want images in the head. Only the things themselves. A universe of objects. So the experience has to be the object itself. But for that to happen the old, reliable reassuring object—the bed I find when I go to my bedroom, the milk carton I know will be there when I open the fridge—has to be broken up and relativized. However, I'm sure the question everybody out there wants to ask you is the same question people used to ask Bishop Berkeley, the guy who said the world was all ideas and the tree didn't exist if no one was there to look at it: What the hell is your famous apple when it's not relative to me?

Manzotti: Tim, please remember: the apple is not relative to *you*. The apple is relative to that object that happens to be your body. *You* are your experience, in this case the apple. But to address your objection: when you're not around for the apple, there is, of course, something still there on the table. But it's not the apple you are familiar with. It is what is there relative to the table, the room, the earth's gravitational field, and so forth.

Parks: But how can there be a *table* if it needs my relative presence to be a table! Or no?

Manzotti: The table needs another object. At least one. After all, what would the velocity of an object be if there were no other objects around? A lonely atom in a void universe would have no velocity whatsoever. The table is surely something for the apple even when you are not there. Likewise, the apple is something for the table. But what they are relative to each other is not what they are relative to your body.

Let me put it another way: in the room we have a whirlwind of physical states. This whirlwind contains a lot more than a human being could ever perceive—atoms, neutrinos, photons, quarks, strings, quantum fields—a huge range of possibilities. When the body comes into the room, its sensory capacities carve out *one possible subset* of that whirlwind. Or, looked at the other way round, one possible set within the whirlwind finds, relative to the body, a suitable causal path along which to roll. So the table and the apple are born! My body brought them into existence in the sense that it *selected* them and *only them* from the whirlwind. Entirely ignoring all kinds of other stuff.

Parks: This whirlwind is disconcerting. As if we lived in chaos. Maybe you should find some more reassuring term. But, summarizing, there is an apple relative to my body, another apple relative to your body, another apple relative to a bat's body, or a bird's, or a fly's, or a possum's, or a microbe's, and so on. Aren't there too many apples? Should we call them apples at all?

Manzotti: Go back to velocities. Each object has infinite velocities. But this does not bother anyone, nor does it disturb our knowledge of physics. As long as you know what reference frame you're in, multiple velocities is not an issue. It's the same with the apple. My apple and your apple are each relative to a different body. So we have relative apples, as we have relative velocities.

Parks: But what about weight, color, size. Are they relative too?

Manzotti: Weight is relative to gravitational pull. Colors vary with light, sensory receptors, surroundings, surfaces, and so on.

Parks: And mass?

Manzotti: Mass is not something we experience. We experience weight. More generally, we perceive an object's resistance to being lifted, pushed, or pulled. I'm not claiming that *all* physical properties are relative, only that the properties *we experience* are relative to our bodies.

Parks: What about size, though? We perceive size, but you're surely not claiming that an object's size changes.

Manzotti: If you insist on a fixed frame of reference, a ruler for example, you could say the size of an object remains the same in relation to the ruler placed alongside it. The apple will always be four inches high and the ruler will always give the same reading. But if you consider the object in relation to the body, obviously it changes with distance.

Parks: Sorry, I've done some homework here. In 1758, David Hume wrote, "The table which we see seems to diminish as we remove farther from it: but the real table, which exists independent of us, suffers no alteration: it was, therefore, nothing but its image, which was present to the mind." Respond in fifty words, please!

Manzotti: Hume was wrong. The real table he had in mind is akin to the pre-1600 notion of absolute speed. Adapting Bertrand Russell's famous comment on the law of causality, we could say Hume's "real" table along with notions of absolute velocity are "a relic of a bygone age, surviving, like the monarchy, only because they are erroneously supposed to do no harm." The table, like any other object, is relative and, as far as we're concerned, relative to our bodies. Since our bodies change and move, the world around us also changes, and so do we, since the thing that is our experience is the world around us.

Parks: Let's say I accept that each of us sees a different relative apple. All the same, only one of us can eat it. Then there's only one apple.

Manzotti: Each of us *is* a different relative apple. Each conscious mind is the collection of objects that exist relative to that object which is our body. Of course, if I eat the apple, it will no longer be part of your experience.

Parks: What about when people die? Their consciousness is gone, but the world remains the same. Surely this proves that experience is a personal thing inside us, not outside, in the world.

Manzotti: Not at all. Imagine you're driving your car straight toward the Leaning Tower of Pisa at fifty miles per hour. Does the tower have a relative velocity with respect to you?

Parks: I would never dream of doing such a thing, but yes, of course. The tower has a relative velocity of exactly fifty miles per hour.

Manzotti: Now, your car is suddenly destroyed. A drone strike takes it out. Why not? Boom. Your car is dead, you are dead. Does the tower still have a relative velocity of fifty miles per hour?

Parks: Obviously not. Or not in relation to my particular car.

Manzotti: So, this property of the tower ceased to exist when the approaching car was eliminated. The moving car brought into existence a world of relative velocities that were neither subjective nor internal to the car. Right? It would be mad to look for the relative velocity of the tower inside the car. There is no emergent relative velocity bubbling up from its carburetors. In the same way, each body brings into existence a world of relative objects, that are, nevertheless, external physical objects. Not things that emerge from your brain, or representations that well up in there. When the body stops working and dies, that world of experience, your consciousness, which is external to your body, ceases to exist as well. But not, of course, the whirlwind it was selected from.

Parks: Essentially, you're turning everything inside out. The experience I thought was inside is outside.

Manzotti: That's the idea. Look at the world, and you'll find yourself. Look inside your experience, and you find   what? The world that surrounds your body.

Parks: Well, of course, we've talked over this many times and over the last year or so I have made a big effort—if only out of curiosity—to stop thinking of my experience as in my head and understand it instead as the very world I move in. And I'm going to confess that to a degree this works and is even cheering. I mean, it's heartening to think that this tree I see, this cup I touch, is not a representation, not the concoction of my neurons, but simply reality, albeit reality relative to my body and my neurons.

However, our sense that consciousness is in the head does not arise, it seems to me, from our immediate sensory experience of the world, but from all the consciousness that has nothing to do with an immediate external object. By which I mean language use, thinking, memories, and so on. And here it gets difficult to see how your approach can work. How can you claim that a dream is not in the head? The eyes are closed. The room is in darkness, and yet maybe I'm in my kayak negotiating an exciting mountain river. This, it seems to me, is where your whole house of cards comes tumbling down.

Manzotti: Okay, dreams it is. And, why not hallucinations, too. We will tackle them. We will consider how the brain-based approach explains them—very easily actually—and then how the Mind-Object Identity Theory explains them. But before we do that, I'm afraid we're going to have to sacrifice one other sacred cow. The now.

Parks: The now?

Manzotti: Just as there is no absolute object, so there is no absolute instant of time that is now. This is crucial.

Parks: Crucial, maybe, but definitely too much for me today. Right now, I need a break.

7. DREAMING OUTSIDE OUR HEADS

Sooner or later any theory of consciousness must address this question: How can it be that during sleep, but very occasionally in waking moments too, we have experiences that have nothing to do with the world immediately around our bodies?

The dominant, "internalist" account of consciousness—based on the assumption that consciousness is generated by neural activity in the brain—has no difficulty in responding to this question. Indeed, it's one of the curiosities of internalism that it is most confident when describing those areas of experience about which we are least certain. The internalists say, If I can have the experience of climbing a snowy mountain on a bright day when I am fast asleep in a dark room, this must mean that the brain can generate experience without contact with external reality. Some internalists draw on dreams and hallucinations to suggest that all experience is no more than a form of "reliable hallucination," a movie in the head with only tenuous relation to the outside world.

By contrast, in our last two conversations, Riccardo Manzotti outlined a radically "externalist" account of consciousness, proposing that our experience is not inside the head at all, but actually identical with the many objects that our bodies and brains carve out from all the atoms, electrons, neutrinos, and photons around us. In short, our consciousness is the world—or the objects—that we experience. There are no manufactured representations of that world or those objects in the head.

Such an approach requires a new notion of what we mean by "an object." An object is not something that exists absolutely, but in relation to the things around it; in our case, in relation to our bodies and brains. So every experience must have an object, simply because the experience is that object. Mental and physical object are one. If I perceive an apple, there is an apple out there that is my perception.

This idea has its attractions, granting an absolute reality to our experience, in the sense that I am the world I call my experience. It is not a hallucination. It really is there, even if it is as I see it only in relation to my body. But what about dreams?

—Tim Parks

Tim Parks: Riccardo, surely the fact that we dream is fatal to your theory. Last night, for example, I dreamed I was swimming across a lake in the dark trying to keep my laptop above water with one hand. Who would ever do such a thing?

Riccardo Manzotti: But you like swimming and lakes and your laptop is probably the single most important object in your ordinary life. In general, however strange a dream narrative may be, I would claim that all the objects encountered in it are things we have already encountered, or amalgams of things we have encountered, in our waking experience.

Parks: I'll have to think about this. Essentially you're turning the Shakespearian adage on its head: dreams are such stuff as ordinary experience is made on. All the objects of daily life turn up there, albeit rearranged.

Manzotti: For the internalists this ordinariness should be surprising, shouldn't it? If dreams offer access to an unconstrained mental world generated within, why is their content so familiar? If color is produced entirely and exclusively in the brain, why don't we ever dream of new colors?

Parks: I suppose the point is that the brain rehashes what we've come across already in the world. The dream is replay.

Manzotti: That's the standard explanation. The brain learns what colors are by interaction with the world, then plays them back in dreams, a remix of the movie in the head. But if, as neuroscientists maintain, colors are properties manufactured in the brain rather than existing in the world, why would the brain need the world to manufacture them? And why in dreams would the brain be constrained by the colors encountered in daily experience?

Parks: Surely the idea is that each experience we have alters the patterning of electrical and chemical activity in the neurons. When we sleep these "patterns" fire off again, giving us experiences we call dreams.

Manzotti: So, all the qualities of our vastly varied experience remain somehow attached to neural traces? If this is the case, then tasting honey affects certain neurons that are now capable of reproducing a honey flavor in your head any time they want. But not a flavor you've never tasted before. Why on earth should that be?

Parks: Riccardo, you've elaborated on this fundamental problem of the internalist account in our previous conversations. But this hardly proves your own position. Insisting that we only dream what we've already experienced doesn't get you out of jail at all. The fact is, eyes closed, I dreamed the lake and the laptop when *no lake or laptop was there*. If, as you claim, every experience is identical with an external object, how can you possibly explain that?

Manzotti: In our last conversation, I warned you that to understand dreams we would have to tackle the question of "the now," or nowness. You say you dreamed the lake and the laptop "when no lake or laptop were there." And you say it with great confidence, as if no one could ever get such things wrong. An apple is there when I can reach out and take it. It's not there when I can't. Case closed.

But is experience really so cut and dried? For example, from our window on the fourth floor here in Milan we can see the Alps fifty miles away. Are the mountain tops *there*, the way the apple is there? I can't touch them from here, can I? Of course, I could set off toward them, but when I get there it will be dark and maybe a snowfall will have utterly changed their appearance. So the mountains are less "there" than the apple.

Parks: But that's merely a question of distance! It's banal.

Manzotti: So what about the fly marching across the table beside the apple? It's right there too, isn't it? But you can't grab it the way you can grab the apple.

Parks: Only because I'm not quick enough.

Manzotti: Actually, no. Our hands can easily move quickly enough to grab a fly. The problem is our perception. As Benjamin Libet showed in experiments made in the 1970s and 1980s, in everyday perception there is a time lapse of around 20 to 300 milliseconds between what you perceive and the moment the perception is available for you to act on it. This is customarily explained by saying that our conscious perception is *in delay* with respect to its external cause, a delay corresponding to the time it takes for the light to travel from the object and then for the signals to pass through the various neuronal synapses. However, there's another explanation that matches the same data just as well: the things our experience is made of are *whenever and wherever they are,* even if the neural activity comes later. Our experience is where and when those things are.

Parks: Sorry, I don't get you. You're now claiming that not only is experience distant from the body, identical with the object, but, as it were, *prior* to the body's reaction to it. Is that it?

Manzotti: Logically, you must agree that if the experience is located at the object and identical with it, it must come *before* its effects on the body, since, following Einstein, everyone knows that distance is also time.

Parks: This is nonsense. How can my present experience—my now—be in the past? Neuronal activity is now. If there's a hundred millisecond lapse between the fly and me, that simply means that I see the fly as it was a moment ago, rather than as it is now.

Manzotti: But if the fly you see is different, even by just a few milliseconds, from the fly you're trying to catch, then we are back with the internalist view that your consciousness is a *representation* and not, as I am claiming, *the object itself,* which is always more or less distant from the body in time and space.

But let's push this conundrum a bit further. Right now, we're at that time of day, a few hours before sunset, when we can see—with a little craning out of the window—both the moon, which is almost full, and the sun. And they both seem present to you, right?

Parks: They *are* present. Everything I experience feels now. Otherwise it wouldn't be experience.

Manzotti: But this sun that feels "now" to you is the sun that shone eight minutes ago. The moon is closer, but it's still more than a second away from the effect it produces in your body. If this were a starry night, the stars composing the Ursa Major would be scattered between 80 and 128 years away.

Parks: But how does that change things? We all know that even the closest stars are light years away.

Manzotti: Have patience and we'll get there. Your now, your experience, is made up of all the things causing an effect in your body, your brain, no matter whether from centuries or milliseconds ago, inches or light-years away. But you don't partition your immediate visual experience into past and present; to you everything, sun, stars, apple, fly, is just now. The present, then, dependent on your neural activity, is what caused that neural activity to occur, however near or far. Your now is not a single point on a linear flow of time. It is spread across time and space. There was never a single instant when the stars of a constellation were in the relation to each other that you experience. Because each star is a different time away from your body. As the fly and the mountain are also different times from your body. So it is only relative to your body that the world, which is your composite experience, manifests itself as it is to you.

Parks: Didn't Einstein start his theory of relativity with a critique of our ordinary ideas of time?

Manzotti: Indeed. Einstein denied that events were "simultaneous" in the sense of being "instantaneous," as Newton thought. There was no absolute clock ticking away to establish time across the universe. Rather, things interacted together in a mesh of causal processes. All I am doing is applying Einstein's insight to my understanding of everyday perception.

Parks: I can see there's a logic in saying that if my experience is located at the objects of perception, and not at my body—spread across space—then it must also be spread across time. I can also see that there's a link between the internalist notion that our conscious experience is located at a single point in space—between the ears and behind the eyes—and the conviction that we are located at a single point in time. In fact, if consciousness is located in that single point between the ears then logically it must be concocted at that point, since any distance implies a time-lapse. The object, which we all know is at least milliseconds away, cannot be identical with the experience in my head. So the experience must be a representation.

Manzotti: Exactly. But let me add that even the internalist approach has difficulty with the idea of time as a point, an instant of nowness. Because the quickest neural processes still require tens of milliseconds to complete as electronic and chemical signals travel back and forth across the meters of neural circuitry packed in our brains. So even for the neuroscientists, the physical stuff that they view as constituting our conscious experience is spread over space and time, albeit tiny spaces. In fact, if we wanted to be really rigorous and consider only what is present at one instant in time, the world as we know it would disappear. Sounds, light, voices, gestures, actions, words, all require a flexible notion of nowness that encompasses more than a single instant. Once you accept that, there's no reason to resist the notion of the "relative now" that I am putting forward.

Parks: Relative in the sense of relative to my body?

Manzotti: Right. For an object to be present simply means that it's *causally present*, it's having an effect on your body, even if it's not actually in the temporal or spatial vicinity.

Parks: Got it. And I think I've got your take on dreams, too! When I dream the lake and the laptop, those real objects are having a causal effect on my brain. They are not stored in the brain, they are really affecting it. Hence they are present. The way a distant constellation is present, even if actually each star is a different number of light-years away from us and they were never, as you said, "simultaneously" in the physical relation to each other in which I see them.

Manzotti: You're jumping the gun. There were other things I wanted to say first. But yes, we're eliminating the naive idea of time as a linear progression of single instants and substituting it with a mesh of causation, just as Einstein did when he explored his theory of relativity.

Parks: But that also means that you're giving dream experiences *the same status* as ordinary waking experiences. They are both the result of the same processes of causation.

Manzotti: Absolutely. And that's how it is, isn't it? When you dream, it's real. You are the objects of your experience. You (and remember, we've said that you are not your body but the experience your body makes possible) were swimming in the lake holding your laptop above the water.

Parks: Desperately anxious that it would fall in. I woke in a sweat.

Manzotti: There you go. Nothing more real than anxiety!

Parks: Very nice. But I'm afraid I don't buy it at all. And I don't buy it for this reason: When I see the constellation arranging itself across the light-years, there is a direct line of vision between its various parts, the stars, and my body. It makes physical sense, *scientific* sense that I see it. But in the dream there is no direct line between the lake and my body.

Manzotti: There's no direct line through space, but there is a causal line. Let's say you visited a lake a year ago. Your body and brain allowed that stretch of water to start a causal process, as a result of which it was part of your visual experience; in short, you saw it. But now let's suppose the causal process *didn't stop at your visual cortex*; it went on to produce a cascade of further effects, beginning a long journey made possible by the unique structure of the brain that allows the causal influence of external events to tick over, as it were, in the background, or as if trapped in a whirlpool, waiting to release its energy when conditions allow.

Parks: I'm sorry but this is beginning to sound rather magical now. And all so convenient for you. I don't get it at all.

Manzotti: Tim, you're a kayaker, aren't you? You paddle your way down wild mountain rivers.

Parks: I do, indeed. Some of the best moments in my life.

Manzotti: But you don't just barrel straight down the river, do you? You take time out here and there. Maybe to wait for others to catch up. How do you do that if the water is moving so fast?

Parks: Easy. Behind each rock, or spur in the bank, there's an eddy. The water gets trapped there, turning round and round. Shoot into the eddy and you're out of the stream.

Manzotti: Great. So now imagine the brain with its billions and billions of neurons, its trillions of neural synapses as a huge collection of eddies, each of them the offshoot of a different original external cause. In everyday perception, the powerful current coming from the proximal surrounding environment is so strong that the eddies don't contribute to the overall flow. Things turn round and round there in a potentially never-ending merry-go-round. Like the Coke can that floats round and round in an eddy. But what happens in a river if the water level falls and the current eases off?

Parks: There's less tension between eddy and mainstream. The waters mingle. The Coke can escapes.

Manzotti: Right. And when you sleep, the flow of immediate experience eases off and the endless eddies can mingle with the flow.

Parks: Nice analogy, if nothing else... Let me sum up. There's an original cause, say a Coke can, that, thanks to its effect on my body and brain, becomes an experience. From then on the Coke can will be present whenever that ongoing effect, in the eddies of my brain, is drawn back into the mainstream. At which point my experience will again be the Coke can, however distant in time and space. So my dreams, but I suspect you're going to say my memories too, are nothing but direct perception of my past.

Manzotti: Careful now! There's no literal Coke can in your brain, nor do the neurons bear the Coke logo, as it were. We just have a causal wave started by the can, and your experience of course is located *at the can*.

Parks: In the past.

Manzotti: I'm not sure it's helpful to say that. Because it implies that you know when your *now* is, and where. But do you? You experience a star as it was a hundred years ago; so when is your experience? You say your experience is at this moment, one hundred years later, when the light from the star reaches you. But why should your experience be at the same *when*, the same moment, as the neural activity that doesn't have anything in common with the star that produced it? Why should experience be located at the point of neural activity when that activity has none of the properties of your experience?

Parks: So, you're claiming that the belief that our experience is where and when neural activity takes place is a mere prejudice?

Manzotti: Right. The star, the apple, the lake, the laptop, the sun, the moon, the distant Alps, the Coke can, and, yes, our friend the fly now buzzing on the windowpane, these are the things that have the properties of our experience. That is as true when we dream as when we are awake. And however great the time gap between their occurrence and the completion of the neural activity that allows them to be present, in consciousness, our experience is always the object itself, whenever and wherever that object is.

Parks: So, our bodies and brains are at the endpoint of a causal process, that can be very brief, or interminably drawn out. But we, our experience, are at the origin of the process, not the end.

Manzotti: Exactly.

Parks: Even if that origin, as with the stars, occurs before I am born.

Manzotti: Before your body was born.

Parks: But wouldn't it be so much easier to think of the experience as the whole process, from origin, through the photons across space and time, to the neurons in the head?

Manzotti: We've been here before. The process is *necessary*, but it does not have the properties of the experience. It supports the experience, but it is not *it*. The only thing that has the properties of the experience is the object itself.

Parks: Enough. I shall now spend a few days dreaming up some serious objections.

8. THE BODY AND US

What is the function of the body in consciousness? Am I my body, or my brain, or a part of my brain? Could I ever exist separately from my body, my consciousness downloaded in a computer, for example, or received into heaven?

So far my dialogues with Riccardo Manzotti have presented two sharply contrasting accounts of consciousness. The standard "internalist" view assumes that conscious perceptions are representations generated by the brain's neurons in response to input from the world without. The radical externalist view—the Mind-Object Theory—put forward by Riccardo suggests that our experience, or perception, is the object perceived. There is no internal representation; body and brain are simply the conditions that allow the world as we know it to manifest itself as it does. And we are the world we experience. Or to put it another way, your experience is you.

Both approaches have enormous implications for a theory of mind. The internalist view tends to align with traditional ideas of the self as an entity located or centered in the head. Though this subject, or self, is obviously dependent on the body for its existence, the internalist model nevertheless holds out the hope that, once mapped and fully understood, the brain could be copied, or its hugely complex electronic and chemical patterning downloaded onto some other "hardware" that would then

*make the subject, the self, immortal. In short, the dominant
internalist position, though differing from the Christian and
Cartesian position in being entirely materialist, nevertheless
allows us to go on thinking of ourselves as at least potentially
separate from the world around us where all is in flux.*

But what about the Mind-Object Theory?

—Tim Parks

Tim Parks: Riccardo, if I accept that when I see an apple, my
experience simply is the apple, and is external to my body, then
we have eliminated the traditional distinction between subject
and object and with it the possibility of any experience that is not
the material world, any interior "mental" existence that could be
separated from it. So where is, or what is, the "I," the subject? And
how are we to think now about the relationship between body,
brain, experience, and self?

Riccardo Manzotti: Let's go back to basics. It's one thing to say,
as I certainly do, that you can't exist without your body and quite
another to say that you are your body. The two claims are different, but traditional materialism has tended to conflate them,
probably because their enemy was the Cartesian and Christian
notion of the immaterial soul. To combat that anti-scientific, religious position, they insisted that the self must, like anything else,
be material; and being material, that it had to be the body.

Parks: Or more specifically, the brain. Since most people would
accept that you could lose a part of your body—an arm, a

leg—without ceasing to be yourself. Whereas you couldn't lose your head.

Manzotti: Of course. You can even have a face transplant these days and still remain the same person. The result is, yes, that since for the scientist whatever exists is *material* of some kind, the subject, the self, has to be material and the brain has always been considered the prime candidate. Hence the absolute determination, over the last century or so, to locate consciousness and the self—some command center we could call the subject—in the brain.

Parks: And you're saying that this is a false assumption. It is possible to look for a material self, without identifying it with the brain.

Manzotti: Absolutely. The alternative to both the material brain and the immaterial spirit is staring us in the face; I mean the external object, all the things, the very physical things, out there in the world that form our experience. Of course, very often, the body itself is, as it were, *part* of the world we experience. The sight of our hands, the feel of our fingers, our face in the mirror, the sensations of walking, running, or simply sitting or breathing. But it's only a part. When we experience our hand holding a tennis racket, both racket and hand are equally our experience and equally us.

Parks: You're going too fast for me. What I was trying to ask is this: Assuming we accept your notion that we are the very world we experience, all the different things we see and hear and smell and feel—how can you construct from this reality the feeling of subjectivity we know in every moment, the impression we have of acting and choosing and planning? How can you explain the way we *identify* ourselves with our bodies, and above all our faces? We have a sense of ourselves. Why aren't we simply a scatter of apples and tables and laptops and trees, walls, doors, floor tiles—of all the things around us?

Manzotti: Now *you're* going too fast. Let's stay with the body a moment. Of course it's absolutely central. There would be no experience without it. Yet *we* are not our bodies. We are something else. Think about it. We do not experience being neurons and blood vessels. We do not experience being bowels and internal organs. Science teaches us that we are made of such stuff, and constantly invites us to contemplate models of our skeletons and innards and so on, and to identify with them. Yet for thousands of years people never thought of themselves like that at all. Because actually our lives are made up of external events, people, objects, landscapes, and of course the body's interaction with these things. What was Homer's experience made of? Chariots, walls, towns, spears, wounds, armors, seas, ships, sails, sacrifices. What was F. Scott Fitzgerald's experience made of? Fast cars, designer clothes, expensive houses, pink cocktails, jazz. As regards their *bodies*, Homer and Fitzgerald were made of the same stuff, neurons and cells. But their experiences, hence their minds, have little in common.

Parks: Well, beautiful women. Fine literature.

Manzotti: Of course. But you take my point. The body's perceptual apparatus, eyes, ears, nervous system, selects which object becomes your experience, carves out a world that is you, but it does not *concoct* this object in the neurons in the brain. The object is out there. Your experience is out there and you with it. The body is a selector and a facilitator, not a host or a container.

Parks: It's true I may not *experience* my neurons, as such, or the cells in my bowels. But I know when I have a stomachache, a headache, and so on. Hunger is an experience. Satiety is an experience. Not all my experience is *out there.*

Manzotti: As we have said, your body is an object too. So of course you perceive it. Your eyes are engaged and affected by light reflecting on the skin of your hands or legs, and you see them. But it's your hands you see, not your eyes.

When you clap your hands you hear their percussion, but you don't hear your ears. And when you feel your bowels, because you've eaten too much, perhaps, that's because they're producing an effect elsewhere in your nervous system. They are *causally* present.

This is always the case with experience of the body. One part of the body perceives another part, but not itself. Notice that we never feel the brain, because there is nothing beyond it, as it were, in the nervous system, that might allow the brain to become manifest to us as an object. No anesthesia is required for the brain itself when a surgeon operates on it, because the brain doesn't feel pain. It allows other objects to exist and become part of our world, within the body and without, but isn't experienced itself.

So the body is part of your experience, just as things in the world outside are part of your experience. And both are you. For perception, there is no "magic threshold," as the philosopher Teed Rockwell put it, between the body and the world. The only thing that makes the body "more important" is that without it there would be no experience at all. And of course it is there all the time. And we protect it as best we can, for obvious reasons.

Parks: Yet if I want I can close my eyes and shut out the whole lot. And I don't feel I exist any less than when my eyes are open. Rather the contrary. In fact, in a dark well-insulated room with no smells, no noise, I will feel *more* myself than ever, freed from external input. Free to think about what I want. In this sense, surely I can be separate from the world. A subject outside the world.

Manzotti: Hold on. In that dark, silent room your immediate experience is darkness and silence, exactly what your surroundings are. Perhaps certain bodily sensations come to the fore. For example, in the pitch dark one often becomes very aware of the need to balance the body, since all reference points are gone. Or in the silence and stillness you experience your breathing, or even your heartbeat. But these experiences are all physical. If your thoughts now move away from the immediate world around you, they will go to previous experiences, people and objects and even

debates, like this one, that again are all part of an external reality. Which is merely to say that when our engagement with our
surroundings is less urgent, when we don't have to interact with
the objects around us, we are prone to perceive our past in various
combinations, something we talked about in our last exchange on
dreams. But the point I am trying to make is that at every single
moment, *you* have to be *something*, an experience identical with
an object, where, by object, I mean something, anything, whether
immediate or distant, in time or space, that is *causally affecting
your body*. An argument you had with your son last night. The
smell of a food you ate on holiday. Whatever. And you *are* that
something, the thing that has the property of your experience.

Parks: You're right of course that when one opens one's eyes
one simply can't avoid seeing things. Right now I have to see the
university office we're sitting in. Nevertheless there's a constant
process of focus, of selection, isn't there? *I, me,* am active in the
composition of that experience. *I* decide what to look at.

Manzotti: Of course. This is true of all experience. When you go
to a party, you don't experience the whole event. You choose who
to talk to, who not to talk to.

Parks: I wish! I always seem to be getting stuck with people I don't
want to talk to at all!

Manzotti: Well, you can't decide who's going to be at the party,
who'll bump into you, or who will be talking to the woman
you're interested in when you try to say hello. But that's not what

I'm talking about. When I say "you choose," I'm simply referring to the fact that your body, with its perceptive apparatus, singles out a subset of the people present, a sub-party in the whole party. That sub-party is *your* party. So your body is the conduit that determines *which* objects in the world are you. Nobody experiences the whole party, or the whole world. Since experience comes from having a body, it can only involve the objects that enter into a causal relationship with that body.

Parks: I get that. But it seems to me you're still avoiding my main question. If I am the world I experience—my party, if you like—what is this sense I have of being a *subject* separate from the world? How can I be both subject and object?

Manzotti: What you call a subject is nothing but a particular combination of objects that are relative to another object, your body. Being a subject means no more than being experience, i.e. a collection of objects, relative to your body. You ask how, if this is the case, the feeling of "subjectivity" can arise. My answer is: thanks to two misconceptions.

Parks: So let's hear them!

Manzotti: First, precisely because the interaction of body and the surrounding world creates a relative world that is unique to that body and to no other, we each experience a world that is different from what others experience. Our world, not theirs. I am near-sighted, you far-sighted. I am on this side of the room, you on that. And so on. We then suppose that the differences between our experiences mean that those experiences are "subjective" and constitute a private and internal realm. In fact, of course, our worlds are not different from the world as a whole in the sense of being concocted in our minds, or woven out of an immaterial mental fabric, *regardless of an external reality*; they are simply the result of the intersection of this unique material body with these unique material circumstances. And, in fact, when our physical faculties and circumstances overlap, mine and yours, so does our experience. All those whose eyesight permits them to get a driver's license stop at red traffic lights. All those with standard auditory apparatus hear the difference between treble and bass. Et cetera.

Parks: It's not so much that we're separate from the world, then, as separate, or distinct, from the world of others and other possible worlds. Aware of looking at this, rather than that.

Manzotti: Right. It's the body that is separate from its surroundings, not *us*. But let me get to the second equally crucial misconception: our tendency to confuse the body with the "person," or the self, when very obviously the self is *not* the body. The body gets sunburned, the self does not. When we are tiny children our parents point at our bodies and exclaim: "What a lovely cuddly dear baby, you are"—at least Italian mothers do!—and of course we believe them. We must be the thing their fingers are pointing at. This identification is then corroborated by what Daniel Dennett famously called "the center of perceptual gravity." Most of our sensory organs are located where our head is: eyes, ears, nose, and mouth are all there, not to mention the fact that from the tactile perspective, the tongue and lips are hugely important. So we feel we must be located where our bodies, and particularly our heads, are. But in scientific terms, this doesn't imply anything. Then there is the whole social aspect; identifying the person with the body is good for tax collectors, police, and statisticians of all kinds, but it hardly amounts to a metaphysical or scientific statement.

Parks: So the combination of an early identity of self and body in childhood, together with an awareness that my experience is unique to me, creates, you claim, the illusion that the self *is* the body, or, since we never see the self, a privileged private space internal to the body, or the head. Whereas, in your view, the self is an ever-expanding accumulation of experiences made up of those external objects, thousands upon thousands of them, relative to our bodies, either immediately present or still causally active on us despite now being in the past or distant from us?

Manzotti: Yes. The self is just as *physical* as the body, and equally important. It exerts its influence through the causal conduit that the body offers; it is that particular world that the body both brings into existence and reacts to. The body is the fulcrum, if you like, but the external object, your experience, all your experience, over the years, is the lever, the self. The lever is only a lever because the fulcrum allows it to be so. But the fulcrum is only a fulcrum because there is the lever.

Parks: I'm a little lost with fulcrums and levers, though I think I've got your central idea. And yet... you still haven't answered my earlier objection. Surely if I can *choose* what to focus on—your face, or the wall behind you—then *I* am neither you nor the wall, but a choosing, discriminating subject separate from both. You have to settle that issue for me: my choice, which guarantees my status as subject, rather than just another object in a deterministic chain of cause and effect.

Manzotti: Tim, we will exercise our free will by leaving free will to a dialogue of its own! Next time.

9. CONSCIOUSNESS: WHO'S AT THE WHEEL?

For any materialist vision of consciousness, the crucial stumbling block is the question of free will. A modern, enlightened person tends to feel that he or she has rejected a mystical, immaterial conception of the eternal soul in exchange for a strictly scientific understanding of consciousness and selfhood—as something created by the billions of neurons in our brains with their trillions of synapses and complex chemical and electrical processes. But the fact of our being entirely material, hence subject to the laws of cause and effect, introduces the concern that our lives might be altogether determined. Is it possible that our experience of decision-making—the impression we have of making choices, indeed of having choices to make, sometimes hard ones—is entirely illusory? Is it possible that a chain of physical events in our bodies and brains must cause us to act in the way we do, whatever our experience of the process may be?

In my conversations with the philosopher Riccardo Manzotti, we have explored his Mind-Object Identity Theory, a hypothesis that shifts the physical location of consciousness away from the brain and its neurons. Rather than representations in the head, Riccardo suggests that our experience is made up of the very world we perceive. But if this is the case, if subject and object are one in experience, does this not make it all the more difficult to explain our impression of free will? Isn't it precisely our moment-by-moment awareness of making decisions that proves that we are separate and sovereign subjects moving in a world of objects that remain quite distinct from us and over which we have an obvious mastery?

—Tim Parks

Tim Parks: Riccardo, how is it that we can constantly decide to do this rather than that, or just to look at this rather than that if, as you suggest, mind and object of perception are one?

Riccardo Manzotti: The question is: When we choose to do something, could we in fact have done otherwise?

Parks: We certainly have that impression. For example, I am quite sure I could have answered differently to your question.

Manzotti: But could you really? If the universe was rewound a moment, given your thoughts, feelings, and circumstances, would you do anything different? And why would you, if you, whatever you are, were exactly the same? Surely if you did something different, then you would *be* different; the thoughts, feelings, etc. wouldn't be the same. So maybe the really pertinent question is: When we choose to do something, *what are we*, what is the thing that is the cause of our actions?

Parks: Well, mainstream science tells us that essentially we are our brains. Didn't Francis Crick say that we are nothing but our neurons and their activity? In his book *The Brain: The Story of You*, David Eagleman claims that, "Who you are depends on what your neurons are up to, moment by moment."

Manzotti: Right. Crick and many other neuroscientists are convinced that we are our neurons and that these neurons, which are of course physical things, somehow make our choices. The problem is that when we use modern microscopes to look at our neurons, we don't find any evidence of this. All we see is a passage of electrical charges and complex chemical changes. Some people no doubt take consolation from the idea that they can blame their gray matter for their sins, as in the past they liked to blame the devil or fate. Eagleman, in the book you mentioned, describes with some satisfaction how a man went on a shooting spree, killing thirteen people, as a direct consequence, he claims, of a small brain tumor "the size of a nickel," which pressed on his amygdala and upset all the neurons there. In this scenario, then, we attribute moral blame to a bunch of cells. But this is hard to square with our actual experience of living and acting in the world. We don't feel an identity with our neurons and we do feel we are responsible for what we do. So, again, the question is: What are we?

Parks: I notice that when *I* say I have a strong instinctive impression of something, you call my experience into question. But when a neuroscientist says we are our neurons, you appeal to instinct and experience to deny it.

Manzotti: Our experience offers a starting point. We have this or that impression, okay, so let's test it scientifically. Crick has neither experience nor science on his side when he claims we are our neurons. Our experience does not offer evidence that this is the case, and despite years of research, it has not been demonstrated.

Parks: I'm sure neuroscientists would disagree with you. For example, didn't Benjamin Libet demonstrate as long ago as the 1980s that our brains anticipate our conscious experience of deciding to do something? When we press a button, for example, there is neural activity tending in that direction as much as a second or two before we can report "deciding to press it."

Manzotti: Absolutely. In fact, recent research by Patrick Haggard and, independently, by John-Dylan Haynes, has confirmed Libet's findings. Well before we are aware of our conscious will, the neurons are busy in that direction. When John suddenly decides to kiss Mary, his brain was actually ahead of the game.

Parks: But surely this confirms the neuroscientist's claim that we are our neurons, since the neurons are calling the shots.

Manzotti: Not at all. You're leaving something rather large out of the equation. In the case of pressing the button, for example, you've forgotten the button. In the case of John's kiss, you've forgotten Mary's lips. You are speaking as if the brain were entirely separate from what is outside our bodies. I hope we've established in our previous conversations that the objects that become our experience are not "absolute" but "relative"; they are as we know them because our body with the causal structure of its perceptive system carves them out thus from the mass of atoms and photons round about. The object, whether it be a button we are about to press or a mouth we are about to kiss, is *relative to our body* and only as such is our experience. We are identical with that experience, not with our bodies or brains.

Parks: I'm sorry, I know we've spent a great deal of time establishing your notion of the identity between object and experience, but I don't see how this can explain how action is initiated. You can't tell me the button decides to be pressed, or not pressed, if it comes to that. As for Mary, she may very well give John a slap on the face when he moves in for the kiss.

Manzotti: Let's take a simpler example, since for the moment we have no idea what pressing that button might lead to, while kissing of course implies another body and another brain.

Parks: I'm all for simple examples.

Manzotti: Okay. When I see an attractive new car and decide to buy it, what is the cause of my action? Couldn't it be the car itself? Why should I introduce an intermediate entity between the car and what my body does? Why not imagine instead a perfectly natural causal chain?

Parks: I really can't see the purchase of a car as simple. This is a big-ticket item. It's a nightmare for me when I have to buy a car. I lose sleep weighing up all the pros and cons.

Manzotti: One doesn't buy a car every day. You're right. Many factors are involved. Your journey to work. Your budget. The kind of roads you use. But all these things are external to your body and, as we said, exist relative to it. Together they make up the composite relative object that is your experience, the car. So why shouldn't it be this car experience, rather than your neurons, that determines your decision? After all, you can't decide to buy the car if it doesn't first exist, if it isn't in some way part of your experience—part of you—even if only through magazines or hearsay.

More generally, what could we mean by *the pronoun I* if not the thing that is the cause of the actions my body initiates? As it happens, neuroscientists agree that I must be the thing that is the cause of my action. And they locate that thing, that cause, in the brain, the neurons. But neurons are *not* the beginning of the causal chain. Their activity is caused by something else: the external world. If we trace any neural activity back, dendrite by dendrite, synapse by synapse, sooner or later we come out of the brain through our sense organs and inevitably find ourselves outside the body, in the world, where our experience is. And we are our experience.

Parks: Well, it's easy to accept that any object I'm attracted to must have some part in my decision to buy it, or grab it. But aren't we merely repeating Steve Jobs's truism that people don't know what they want until you show it to them?

Manzotti: No. We're going a step further. We're saying people don't know what they *are* until you show it to them. Once we are shown the iPhone, say, once our body with its sense apparatus carves out that fantastic object, we are changed. We become the object our senses allow to exist, in this case the phone. So it is with all our goals. Showing people things is very powerful. Hence the world of advertising!

Parks: Still, people make different decisions, don't they? While half the world was lining up to grab an iPhone, I was holding off. I deliberately chose *not* to buy. Surely that's because my brain was calling the shots, not the object, which you claim is one with experience. And, to go back to Libet, as I recall, he remarked that the brain could always change its position at the last second or millisecond. It can change course. So a decision is being made. The neurons are busy.

Manzotti: What would that late minute change of neural activity be caused by? John's on the brink of kissing Mary when he catches a glimpse of Mary's husband coming out of the bank. Bad news. Or maybe he holds back because something in his unique past is still affecting his actions right now, some cautionary tale he heard when he was a child. For me this past experience is, again, an object, a piece of the external world made possible thanks to your body. So one piece of world contradicts another. About to buy the world's most exciting car, I am suddenly aware of my latest bank statement. It is still causally active in my head, part of my immediate experience. In your case, something in your past resisted the iPhone. It had nothing to do with your brain calling the shots. You are the cause of your actions and inactions; but that "you" is not an invisible ghost in your brain but the relative world your body has brought into being.

Parks: I guess your point is that we always do what we want at that instant of doing, perhaps despite other pressures, other experiences, that in another moment might dominate. And if we didn't want it, we wouldn't do it.

Manzotti: Right, but let's not imagine this exempts us from our responsibilities. Rather, it reveals what we really are. We are the causes of the things we do, and our actions are the effects of the things we are. We are that collection of experiences/objects that, given the prevailing circumstances, do what we do. If we lie, we are liars. If we fight, we are fighters. If we love, we are lovers. The cause is defined by its effects. "Ye shall know them by their fruits."

Parks: Matthew 7:16! My biblical childhood is still causally active, it seems.

Manzotti: Ha! What a memory! But notice the element of *necessity.* You cannot *not* remember that biblical reference, you are simply obliged to produce that fruit, being who you are, the evangelical clergyman's son. And notice that this kind of "determinism" is not disturbing to you, because you do indeed identify with the person who had the biblical education and produces the reference. But if I tell you your neurons decided to remember that, then you worry that you are a puppet in the hands of something alien, some strange gray matter, because you don't identify with your neurons. And rightly so, however important they may be! You are your experience, by which I mean the world as made possible by your body.

Parks: Where does that leave the concept of free will?

Manzotti: We often confuse freedom with arbitrariness, as though freedom were tantamount to doing something in a random way. But we are only really free, or rather we savor our freedom, when what we do is the necessary expression of what we are. Someone choosing to come out as gay doesn't do it lightly. They do it because they feel they *have* to. They have reached a point where there is no alternative. Yet, it is in this necessity to come out that freedom is achieved. Freedom is to be one and the same with oneself, with the accumulation of one's world of experience. This is what we mean by identity.

Parks: We still have the issue that Libet raised, that neural activity anticipates conscious decision.

Manzotti: Libet's work, and indeed all of neuroscience, fits perfectly with the model I'm suggesting. Imagine an action as a dam bursting and leading to a new object, a flood. Obviously, the cause of that flood is not uniquely the dam breaking, but the heavy rain, or perhaps years of erosion and poor maintenance, that came before it. In the brain our neurons are affected by all kinds of external causes over time. Hence many separate experiences are building up a readiness to act. But it's only when the dam bursts, or the body acts, that the cumulative effect results in us making a decision. At that point what we do seals what we are.

Parks: As always you are moving experience outside the head. So here, you are seeing the experience of decision making, which I presume you don't deny, not as a negotiation or even conflict between "warring networks of neurons," as Eagleman describes it, but as a coming together of different external objects/experiences pushing us in different ways.

Manzotti: No, not quite! Those objects experiences are not pushing us, they *are* us. They are pushing *our body*. It is the relative car that is your experience, not an absolute car, that finally moves the hand to the wallet. It is the world relative to our body, our perceptive faculties and accumulated experience, that is the cause of our action. The situation is complex and can't just be described as external factors determining our action.

Parks: I see what you're saying: my experience, which is none other than the accumulation of all the objects my body has encountered, eventually determines my actions. But I'm not altogether convinced. And my problem is this: not only do I have the impression of making decisions, cogitating, not just acting, but I also believe that I "organize" experience. That I see the world *in a certain way.* I hold a system of political opinions, of aesthetic preferences, and so on. So I *feel* that rather than being a world of objects coming together over time to determine an action, I have an inner world that determines how I organize the outer world. I don't just act *as consequence*; I decide how to act, coherently.

Manzotti: Let me offer an analogy to suggest the fallacy behind your conception. We'll stay with cars. When you drive you turn the steering wheel and, thanks to a complex yet easily understandable coupling of cogs and drive shafts, the vehicle's front wheels turn accordingly. Is there anything mysterious between the steering wheel and the two wheels that turn? No. Just a chain of cause and effect such that given the turn of the driving wheel, the front wheels have to turn.

Okay, now imagine an infinitely more complex object, a human body. The world acts on the body, but before the body is going to translate that cause into an effect, an action, a simply enormous, though of course necessarily finite, number of causal events may take place, inside the body and outside. What's more, unlike the car, which is a fixed object when it comes out of the factory, your wonderful body can change in response to the world, it is teleologically open—so that, to give the simplest example, when you see a face a second time, the experience is different from the first time, because the first experience is still causally active in your brain, hence we have the sensation of recognition. So with this fantastically complex object, the body, we cannot conceive the whole causal chain that precedes an action (this was a favorite observation of Spinoza's) and hence we cannot predict what action will be taken. As a result of this conceptual impossibility, we slip into the habit of inventing an *intermediate entity*, the self, to which we attribute a causal power. We say that I, or my *self*, caused this to happen. But as David Hume said, we never meet or see a *self*; we meet ideas, or, as I would say, objects. The self, this elusive intermediate entity that initiates action, is a shortcut, an invention, a convenient narrative to explain our complex experience.

Parks: Enough. Rather than saying I can agree with this, I'm going to wait a day or two and see what causal effect your arguments have on my now tired brain. But, to be predictable, I want to close with a challenge. You have constantly claimed that the internalist view that consciousness is neural activity has not been scientifically demonstrated. Well, can *you* demonstrate your externalist view of consciousness, *scientifically*? Are there experiments that would prove your position?

Manzotti: Indeed, there are. And since, as we have seen, I am a person who loves challenges, at least of this kind, for our next conversation I will devise some experiments, which, if undertaken, will prove or disprove my hypothesis.

10. A TEST FOR CONSCIOUSNESS?

Will we ever really know what, or even where, consciousness is? Is there any way to get at it scientifically, conclusively? Week by week we hear claims from neuroscientists that would appear to confirm the prevailing "internalist" view of consciousness. If the brain creates a representation in our heads of the world around us through the firing of neurons, the argument goes, then we can identify neural activity that corresponds to particular aspects of consciousness. They tell us that if this part of the brain is damaged it will affect our eyesight. If that part suffers, we will have difficulty moving through space. They show us images based on scans of electrical and chemical activity in the brain and how those images change when our experience changes. Yet there has been no progress in bridging the gap between this activity in the brain and the nature of our experience, the richness of our sensations of color, sound, touch, motion, or simply awareness.

How, then, can the internalist theory be tested and demonstrated scientifically? Will it ever really be possible to prove beyond all doubt that this neural activity is our experience?

And if that can't be done, is there any proof for an alternative account of consciousness? What about the hypothesis that Riccardo Manzotti has been setting out in these dialogues, that consciousness is actually external to the body? Are there any scientific experiments that could settle this debate?

—Tim Parks

Tim Parks: Riccardo, let me start with a very simple experiment, something anyone can try, that seems very much in favor of the internalists. When we look intensely at a field of red color and then shift our eyes to a white or gray surface, we see, admittedly only for a few seconds, but nevertheless very distinctly, an area of green. Since it is clear to anyone who has not been looking at red that there is no green on this white background, is it not evident that colors are generated in the brain?

Riccardo Manzotti: Well, first, you don't see green but cyan, a greenish-blue color.

Parks: Who cares! Surely the only thing that matters is that one is seeing a color that isn't there.

Manzotti: I care, we should all care. When doing science we must be precise. It's actually rather extraordinary that in current textbooks and even in scientific papers people are still claiming one sees a green image after looking intently at red.

Parks: But...

Manzotti: It's more important than you think. Let's put in the colors right here for people to see and have them make up their own minds.

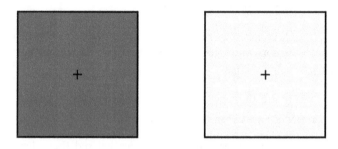

So, readers should stare at the red square for at least twenty seconds—if they're using a small screen, they'll need to get right up close—then move their eyes and look steadily at the light gray, whitish square, where they will now see a color afterimage. But what exactly?

Color A, or color B? If you are a standard color perceiver, a tri-chromat, what you have just seen is much closer to color A than B, that is, to cyan rather than green.

Parks: Okay, it works for me. And so?

Manzotti: Well, white, as you know, or this light gray, is made up of all the colors. And it just so happens that if we take the red out of the white, we're left with cyan. Not green.

Parks: But still, the paper is white, or grayish, not cyan. At least to anyone who hasn't been staring at red.

Manzotti: If you had stared at green rather than red, then when you turned to the white you would see white minus green, which is magenta. And if you stared at red and then looked at a field of yellow rather than white you would see a green afterimage, which is yellow minus red.

Parks: Ah. What you're saying is that what we see is dependent on what's out there.

Manzotti: Right. And we can *predict* what we're going to see. Staring at an intense color, the eye experiences something called chromatic fatigue. It becomes briefly blind to that color. So when it turns to look elsewhere, for a few seconds it does not pick up the color it's blind to. Turning to white after looking at red, you see the cyan in the white. Then white takes over again.

Parks: So I'm seeing something that's really there.

Manzotti: You are. That's why it matters that we establish the exact color we're seeing. Because it's not produced in the head. It depends on what's out there. It *is* what's out there, for your altered perceptive faculties. And before we move on, let me just say that this is a classic example of how an orthodoxy—in this case the idea that experience, and in particular color, is all generated in the brain—leads to some sloppy science and even a denial of what anyone can go and check for themselves.

Parks: Let's see if I can do better with my next challenge. Internalists often mention Wilder Penfield's experiments. He managed to get people to have hallucinations by stimulating parts of their brains electrically during open brain surgery. Other neuroscientists have even managed to relate the stimulation of a particular neuron to "seeing" a particular face, obviously in the absence of that face. Again, this suggests that experience is generated by the brain; we don't need the world around to see something.

Manzotti: Have you checked out the hallucinations Penfield reports?

Parks: No.

Manzotti: They are all rather everyday ordinary experiences. Seeing one's wife entering the room. Hearing a friend's voice.

Parks: And so?

Manzotti: Well, if experience were actually generated freely by the brain, isn't it odd that it remains so strictly tied to the world? Why no colors that have never been seen before? Sounds never heard in reality? Why no experiences that clearly have nothing to do with the outer world? Even when we dream, we are aware that the bizarre aspects of dreams are due to their superimposition or mixing of different elements of known experience. An elephant that's pink, or green. A dog that can talk. Whatever.

Parks: But surely the point is that we're seeing something that's not there.

Manzotti: Tim, we discussed this in our conversation on dreams. The question of what's "there" or what's "now" is complex. The objects that make up our experience can be milliseconds or years away from our bodies. Photons take time to travel, neurons take time to send electrical signals. We have already suggested that although ongoing ordinary experience of the world follows a privileged neural path that makes it possible for the body to deal with phenomena immediately around it, there are also other paths, eddies as it were, where neural activity mills, or is somehow delayed, then released later in dreams, or when a surgeon stimulates a part of the brain electrically. But this does not mean the brain is *creating* experience.

Parks: I'm not entirely convinced by this. You can't prove, scientifically, this idea of experience being buffered or delayed in neural eddies.

Manzotti: At this stage, no. Neuroscientists can't disprove it, or prove that the experience is "generated" in the head. But let's remember, we do science by forming a hypothesis, making predictions in line with that hypothesis, and inventing experiments that prove or disprove the hypothesis.

Parks: So how would that work in the case of consciousness?

Manzotti: Hypothesis: all our experience is made of physical things that have had some causal relationship with our bodies. In fact, if it could be demonstrated that someone has had an experience made up of elements that were never causally related to his or her body, my theory would collapse and—

Parks: Sorry. What about the congenitally blind painter—Turkish, I think—who claims to see colors in his mind?

Manzotti: Esref Armagan. Okay. He was born with no eyes. However, he has spent all his life among people who talk about color and he refers to color with the common terms: the sky is blue, the grass is green, etc. But the colors have to be chosen for him when he paints them. He can't see them, so it's impossible for us to know what it means when he says he experiences them. There are many cases of congenitally blind people writing about color, but they usually admit these are simply words they learned. If color was concocted in their heads, without any contact with the outer world, why would they ascribe the right colors to the right objects, as it were, having never seen those objects?

Parks: I can see we're not going to get very far with this. Your general prediction is that every experience will be traceable back to an actual physical property in the world. But when it comes to fleeting feelings and intuitions, any such tracing back becomes extremely complicated. And I want to be brutally definite. Can you invent a clear and concrete experiment and predict an outcome of that experiment that would prove your position? Accepting, of course, that if the outcome is different, you are wrong.

Manzotti: Yes. Let me propose two. Neither is easy, but then again neither is impossible, and both are certainly easier than much of what neuroscience gets up to these days. The first requires a little surgery and a willing guinea pig.

Parks: Yourself?

Manzotti: I'm up for it, yes. Though no doubt some people will raise ethical objections. So, take an afferent nerve from a part...

Parks: What is an afferent nerve?

Manzotti: Simply a bundle of axons carrying an electrical impulse, or action potential, from an external physical phenomenon to the central nervous system. For example, mechanoreceptors are cells that respond to mechanical forces, such as pressure or distortion. They generate action potentials that head off towards the brain via the spinal cord. They allow the external world to be the cause of effects in the brain.

Parks: Okay.

Manzotti: Take an afferent nerve from a part of the body that is not of crucial importance, for instance a tactile nerve in the back. Then connect it to a transducer...

Parks: Explain.

Manzotti: A transducer is a device that picks up a phenomenon and transforms it into an electrical impulse. For example, artificial retinas and artificial cochleas are transducers, picking up visual and auditory phenomena. Connected to nerves in the eye or ear they offer forms of sight and hearing.

Parks: And what's the phenomenon that the transducer in this experiment picks up? The one we're going to attach to the nerve in your back.

Manzotti: Well, it has to be a transducer for a phenomenon human beings cannot pick up with their bodies. Ultrasound, infrared, electromagnetic fields. Let's say infrared. After all, some species of snakes experience infrared.

Parks: We take the nerve in your back and hook it up to an infrared transducer. Your prediction?

Manzotti: Since my hypothesis is that experience is not *created* in the brain but selected by the brain and the body in the external world, it follows that if we extend the mechanisms of selection, we should be able to extend our experience accordingly. So I predict that as soon as that external phenomenon—in this case infrared—becomes able, through the transducer connected to the afferent nerve, to affect what is going on in my brain, I will begin to perceive the additional external phenomenon. I will have an experience of infrared if only because infrared is now causally connected to my brain.

Parks: This sounds a bit like those attempts to convey visual information through tactile stimulators attached to the back of a blind person. A camera—or visual transducer, I suppose you'd say—sends signals to a sort of plate placed on the back, and the person then learns to interpret the signals visually.

Manzotti: Right. But there are two important differences. First, those systems are not directly attached to the nerves. Second, the point of that research is to allow a person who is blind, but was once able to see, to learn a skill, that is, to respond appropriately to a new kind of visual stimuli—something he or she has done in the past reacting to stimuli from the eyes before he or she became blind. In my experiment, the transducer is fixed directly to the nerve, which puts the body in causal contact with a new phenomenon, not something previously experienced.

Parks: So, we do the experiment, and either you have a new experience, which is an awareness of infrared, or you don't. But couldn't the internalists claim that the nerve was, yes, stimulated from without, but that nevertheless what is experienced is experienced within, and is a representation of infrared, not the phenomenon itself?

Manzotti: Ha! They could. We would have established a need for the outside world to have the experience, but not the location of the experience.

Parks: So you're only halfway there, or not even.

Manzotti: I said there were two experiments and the second attempts to deal with this objection. The idea this time is to prove that it is possible to have different experiences with the exact same neuronal activity. And the experiences would be different because the external world would be different.

Parks: How on earth are you going to do that?

Manzotti: First we need some optical reversing, or inverting, goggles, the kind that make everything look upside down. We know from previous experiments that if you wear the goggles continuously for a few days you adapt and your perception adjusts. You see things the right way up, the way they are, despite the goggles. Right? So, in this experiment, before giving a subject the goggles we present him with a simple visual stimulus, say, a big capital T. Then after he has worn the goggles a few days and adapted to them, we present him the same stimulus, but inverted—an upside-down capital T.

Parks: I'm getting confused. Why?

Manzotti: Well, at this point we have a double inversion: the inverted T with the inverting goggles will cause the viewer the exact same retinal activity he had previously when there was an upright T without the goggles.

Parks: Got it. We've created the same retinal activity with different stimuli.

Manzotti: Right. And my prediction is that despite the retinal activity being the same, the viewer will see the stimulus upside down, *as it really is.*

Parks: Because he's adapted to the goggles. Cunning.

Manzotti: Naturally, we would record the neural activity in both cases using a high-res fMRI (functional magnetic resonance imaging). Here I'm predicting that the cascade of neural activity in the cortical area would be the same, while the experiences, as we've said, would be different and, crucially, correct, on both occasions. This, I think, would demonstrate that the experience is not a neural representation, not in the head, since in the head we have the same activity on both occasions, while the experience is different. Therefore, the experience must exist outside the brain.

Parks: Wait a minute! Wouldn't the adaptation process that the wearer has gone through produce some variation in brain activity, and wouldn't it be that variation that accounts for the different experience?

Manzotti: Yes and no. First, I should say that we'll be recording neural activity related to visual stimuli, the way neuroscientists do when they establish neural correlates for visual experience. Haynes and Rees, for example, in 2006 succeeded in matching specific brain activity with a specific visual experience. More remarkably, in 2011 Nishimoto managed to reconstruct the external visual stimuli that volunteers were responding to on the basis of their brain activity.

In light of these results, then, you might suppose that the adaptation that occurs when someone wears reversing goggles is the result of an inversion that takes place inside the brain. Yet we have no indication that anything of the kind takes place. It's worth remembering that ever since the early 1600s, when Kepler did his work on human vision, scientists and philosophers have been puzzled by the optical inversion that occurs inside the retina and have looked for some corresponding re-inversion in the brain. Nothing has ever been found. As to adaptation to inverting goggles, evidence collected by Linden and Kallenbach in 1999 suggests that no change occurs in the orientation of neural activity in the visual cortex. Of course, one could always object that current brain imaging techniques have their limitations and that there may be hidden neural activities not yet observed, but the burden of proof would then be on the internalists to find such activity. This is an empirical question and needs to be settled empirically, not on the basis of prejudice or dogma.

Parks: Coming at this from another angle, don't we already know that the same type of neural firing along a single axon can be correlated to different senses? In which case, even assuming your experiment works, would it really be such a revolutionary result?

Manzotti: You're right, yes. And we also know that the same byte of memory can have different meanings, and again that the primary auditory cortex and the primary visual cortex have very similar structures with similar neural activity, yet one correlates to auditory experiences and the other to visual experiences. The point of my experiment is to create such a clear-cut situation that scientists would have to consider the obvious conclusion from all this data: that the experience is not located in the brain, but in the truly different phenomenon outside.

Parks: But do you believe that either of your experiments will be carried out in the near future?

Manzotti: At present we are stuck in a dead end where the orthodoxy, internalism, is entirely dominant, but no progress is being made as to the nature of consciousness for the simple reason that, as we showed in our earlier dialogues, this orthodoxy makes no sense at all. Rather than doing any real science, we are hearing fantasies about downloading consciousness onto computers and the like. Perhaps in our next conversation we could consider this state of affairs and challenge internalists to disprove the hypothesis I have put forward.

Parks: By all means, let's see where everyone stands and where they think they're moving.

11. THE HARDENING OF CONSCIOUSNESS

How much of our current worldview, our social organization, our collective psychology, or simply our attitude toward life, depends on how we understand consciousness? The dominant view, which assumes that all our conscious experience is an internal, largely concocted representation of an unknowable outside world, underwrites a number of assumptions: perhaps most importantly, that the human subject is radically split from the object, hence quite autonomous; and again that, unable to perceive the world "as it is," we need science to give us any solid facts we may have.

—Tim Parks

Tim Parks: Riccardo, today I want to take time out from the further development of your hypothesis—that conscious experience is identical with that part of reality that our bodies are able, as it were, to pick up—to focus on the present state of the consciousness debate. In our last conversation, you accused the status quo of being an orthodoxy that does not bear examination and that borders on a religious faith upheld by a collective act of wishful thinking. Can you justify these accusations?

Riccardo Manzotti: To understand the present impasse in this debate, we'll have to focus on the man who more than any other has determined the way in which we think about consciousness for the last twenty years, David Chalmers.

Parks: Okay: Australian philosopher, born 1966, presently at NYU, who in 1996 came out with the famous, endlessly quoted definition of consciousness as "the hard problem." I'd have thought you would be thanking him for bringing the issue to the fore.

Manzotti: Perhaps. Though, as Galen Strawson has exhaustively shown in a fine article in the *Times Literary Supplement* in 2015, consciousness was never not to the fore. What matters for us, though, is that in his book *The Conscious Mind: In Search of a Fundamental Theory* (1996), Chalmers laid out the terms of the consciousness debate in a way that simultaneously excited everyone while more or less guaranteeing that no progress would be made.

Parks: Quite an achievement. To be honest, I can't recall ever hearing anything negative about Chalmers's work. I had the impression that he was universally appreciated for his broad understanding of the issues and commendably affable willingness to consider different approaches. Please tell us exactly what his position is.

Manzotti: I'm afraid that's not something I or anyone else can easily do. There is no "exact" position, which is precisely why he does not excite much criticism or real debate. Over the last twenty years Chalmers has dabbled with panpsychism, dualism, emergentism, physicalism, Russellian monism, and even computationalism.

Parks: That's a lot of -isms. But there must be something specific that you object to. You can't complain if someone offers an overview of various lines of thought.

Manzotti: Let me reply with a few words from the great philosopher Alfred North Whitehead: "When you are criticizing the philosophy of an epoch, do not chiefly direct your attention to those intellectual positions, which its exponents feel it necessary explicitly to defend. There will be some fundamental assumptions, which adherents of all the variant systems within the epoch unconsciously presuppose. Such assumptions appear so obvious that people do not know that they are assuming because no other way of putting things has ever occurred to them." Essentially, when Chalmers so dramatically announced "the hard problem," insisting that we had no solution to the question of consciousness, he simultaneously assumed that the constraints governing any enquiry into it were already well defined and unassailable.

Parks: Well, there must be constraints, surely.

Manzotti: Of course. But when you're making no progress, it's important to go back and examine them from time to time. After that famous 1996 conference Chalmers was given huge credit by the intellectual community. But instead of moving us forward, he has kept us in a stalemate.

Parks: So what are these rules, assumptions, or constraints that Chalmers subscribes to?

Manzotti: I'll name three, though they're all connected:

1. Consciousness is invisible to scientific instrumentation; hence,

2. Consciousness is a special phenomenon governed by its own special laws; hence,

3. It will take a great deal of time and money to fathom these special laws, but if you trust us scientists, we will get there in the end.

Parks: Come on! Your third point can't possibly be a part of a philosophical debate.

Manzotti: I didn't say debate, I said assumptions. Point three is a consequence of points one and two and extremely attractive to any scientific community on the lookout for funding.

Parks: So, within these constraints—an assumption of the invisibility of consciousness to scientific instruments and its consequent special status in nature—and accepting that he has no "exact position," what are Chalmers's crucial ideas?

Manzotti: The idea that conditions everything else is that we can and must distinguish between consciousness and the physical world.

Parks: Please spell this out clearly. I feel we've arrived at something important.

Manzotti: Chalmers, and the status quo in general, split the world in two: consciousness is invisible to observation and measurement, it is *qualitative*; the physical world, on the other hand (which includes the brain itself), is measurable, observable, and *quantitative*. Chalmers called the invisible part the "phenomenal mind" and the visible, or measurable, the "cognitive mind."

Parks: I don't quite get this. He's split the world but both sides of the split are "mind"?

Manzotti: By the "phenomenal mind" he means our consciousness, our experience, our feelings, and so on. By the cognitive mind he means our interaction with the world, the functional machinery of the brain. A ball comes toward me and my hand reaches to catch it, for example. The brain does what it needs to. That's cognitive. But my awareness of the ball-catching experience, which is my conscious experience, is phenomenal.

Parks: Chalmers separated the two.

Manzotti: That was the central idea. Doing so, he made Thomas Nagel's claim that it would be possible to know every detail of the physical world while still not knowing what it is like to be conscious into a dogma. The same divide has been echoed in one way or another by most philosophers and scientists. Ned Block contrasted "access consciousness" with "phenomenal consciousness." Stevan Harnad distinguished between functioning and feeling. So, just as consciousness was placed in the spotlight, it was also removed from the reach of science. Consciousness and cognition emerged as two separate fields of inquiry; you could happily look at one while ignoring the other.

Parks: Isn't this pretty much the same divide made by Galileo and Descartes? I mean Galileo's claim that "tastes, smells, colors exist only in the sensitive body" while "quantities, numbers, and relations" exist in the physical world.

Manzotti: Of course, and Chalmers on a number of occasions has espoused dualist positions, the idea that the world is divided into separate "realms of reality."

Parks: Example?

Manzotti: Well, in his 2014 TED Talk, which you can find online, he says: "Right now you have a movie playing inside your head. It's an amazing movie, with 3D, smell, taste, touch, a sense of body, pain, hunger, emotions, memories, and a constant voice-over narrative. At the heart of this movie is you, experiencing this, directly. This movie is your stream of consciousness, experience of the mind and the world."

Parks: The world is going on in our heads, not out there.

Manzotti: Right. This is Cartesianism in modern terms. We even have a mysterious "you" watching the movie, a sort of updated homunculus, the little guy inside enjoying the show.

Parks: Not exactly. He says, "at the heart of the movie is you, experiencing it *directly*."

Manzotti: If you are experiencing the movie then you are separate from it, and the only way to experience movies is to watch them. And what you are watching, notice, is not the world, but a film. This is Descartes's theater or Plato's shadows in the cave all over again. But it's hardly worth trying to nail Chalmers down on this; there are no silver screens in the brain and no little folk lapping up the action. What matters is the consequence of this approach: the philosopher is happy to have cordoned off a realm that scientists can't touch—consciousness. The scientist is happy that with consciousness removed from the scene he can do his neuroscience without being obliged to find evidence for the philosopher's claims. Meanwhile, the layman is flattered by the notion of a mental inner world to which he alone has access, as if our every waking moment wasn't largely governed by the world our bodies move in. Everyone is happy, but no progress can be made.

Parks: Does Chalmers have no defense against your objections?

Manzotti: Originally, he put forward something he calls "the dual aspect theory of information." According to this view, information has a double nature (an idea that has been recently revamped by Giulio Tononi). On the one hand, it is a sequence of bits without qualities, the on/off setting of elements arranged in sequences that we're all familiar with. On the other, it can metamorphose, in our brains, into colors, smells, sounds, feelings, and so on—the rich nature of our lives. So again, he contrasts a causally effective realm that science can know and a phenomenal mental realm it can't.

Parks: So how does this double nature of information come about?

Manzotti: The theory doesn't say.

Parks: There must be a hypothesis.

Manzotti: The idea is that if you have enough bits, vast quantities of information, such as might be produced by the 85 billion neurons in the brain with their trillions of connections, and if you organize them in a particular way, then a critical point will be reached at which the transformation occurs: bits become qualia, which is to say experience, colors, smells, and so on.

Parks: Why would that happen?

Manzotti: The theory doesn't say.

Parks: I suppose the point is that our computers are filled with images and music that have rich qualities created from bits. So why not our brains?

Manzotti: Tim, you disappoint me. We discussed this in an earlier dialogue. Computers are ingenious devices that exploit electronic states to control other devices—screens and loudspeakers—that make images and sounds. There are no images or sounds literally *inside* your computer, the way there are coins in your pocket. Information does not exist as a physical substance. It is a word we use, a form of shorthand if you like, to describe a process that allows a message to go from sender to receiver. However, in this theory, information, understood as the multiplication and arrangement of bits, at a certain point metamorphoses into conscious experience.

Parks: Rather as if information suddenly became a form of soul or self.

Manzotti: No comment.

Parks: What research is being done to verify this hypothesis?

Manzotti: The idea is to develop more and more complex computers with numbers and arrangements of connections similar to those of the brain, until consciousness "emerges." But no one will take the risk of making a clear prediction that could be tested, so it's hard to say at what point the approach could ever be proved wrong. When nothing happens, you just keep adding more computing power.

Parks: Time and money... But what about the brain itself? I know that Chalmers has done a lot of work on the philosophical aspects of the neural correlates of consciousness, the neural activity that corresponds to our various experiences.

Manzotti: Sure. Underpinning the double aspect theory of information, we have the assumption, subscribed to by many, that neurons "produce" consciousness.

Parks: Tell me more about this.

Manzotti: That's it. There isn't any more. It's an assumption.

Parks: Oh, come on. The wonderful maps of the brain, the correlation between these particular neurons and this particular aspect of experience. It's a sophisticated enterprise.

Manzotti: No one has more admiration than myself for the extraordinary research done to explore the brain and its immensely complex activities. Extremely sophisticated tools have been developed and used with great ingenuity and patience. However, the essential underlying idea here is simply that *neurons produce consciousness*. It's as crude as that. We are simply asking the brain to do what the soul once did. Of course, what neuroscience has actually shown is how neurons consume chemicals, absorb other chemicals and release them, produce and fire off electrical charges, and so on. In many situations such activities occur in strict relation to certain experiences we have. But then, so do the activities of many other cells in the body. And so do the external things we experience.

Parks: So advocates of this idea have no understanding as to why electricity and chemicals should have or produce the qualities of experience?

Manzotti: None at all.

Parks: A dualist water-into-wine situation again.

Manzotti: Not exactly. Water is a real liquid; wine is a real liquid. You kind of figure there might be some trick to do this. The relation between neuronal activity and, say, the sound of a violin seems a tougher proposition altogether.

Parks: Essentially, then, in Chalmers's framework consciousness remains a step beyond physical reality and as a result awareness, selfhood, individual experience, whatever, are all quite separate from the world.

Manzotti: An ordinary man, uncontaminated by this debate, will very likely suppose that his experience is part of the world, in that it is affected at every moment by the world—his body, the environment—and affects the world in turn when he responds to it. But many scientists and philosophers have tried to show that in fact consciousness is, as they say, *epiphenomenal*, meaning its presence or absence is immaterial to the physical world and to what our body does.

Parks: I suppose that's why philosophers came up with the conundrum they call "the philosophical zombie," the idea that there could be people, beings, who behave exactly as we do but lack consciousness. They function, but they don't feel.

Manzotti: Yes. Chalmers himself put the zombie at the center of the debate. Because once you accept that such a thing is logically possible, you accept the split between experience and the physical world, which become incommensurable. Consciousness isn't necessary to describe human behavior, nor can you get at it from outside in any way, so no progress in understanding what it is can ever be made.

Parks: But if consciousness isn't strictly necessary and doesn't connect with the world, why would it have developed, in evolutionary terms? Why would natural selection have taken us that way?

Manzotti: Quite.

Parks: As you said earlier, though, Chalmers has considered many other approaches.

Manzotti: He has been very open. Yet, just as when a writer establishes a powerful new style that others follow, each with their different agendas, so the options Chalmers has considered—panpsychism, for example, or Russellian monism—have all been drawn into the characteristic Chalmersesque zeitgeist.

Parks: Can you, in closing, offer some summary of what is at stake here, some urgent reason why this debate must be moved on?

Manzotti: Man has always liked to think of himself as being at the center of the universe, a special being. Any science that suggests he isn't has always been resisted, from Copernicus's demonstration that the earth moved around the sun, and on through all those discoveries that eroded Man's claim to special status: evolution, genetics, and so on. In declaring consciousness the "hard problem," something extraordinary, and separating it from the rest of the physical world, Chalmers and others cast the debate in an anti-Copernican frame, preserving the notion that human consciousness exists in a special and, it is always implied, superior realm. The collective hubris that derives from this is all too evident and damaging. We should get it straight once for all: there are no hard problems in nature, only natural problems. And we are part of nature.

12. CONSCIOUSNESS: AN OBJECT LESSON

What is "an object" in the end? And what is "the world" that these objects make up?

When we talk about consciousness, we rarely discuss ordinary physics, which we assume science has long since understood: objects are composed of atoms; they exist entirely separate from ourselves, and can be measured and manipulated in all kinds of ways. It's also clear, however, that this idea of the physical world works only if we suppose that consciousness—our experience of that world—is distinct and apart from it; objects exist first outside us, then in a secondary, shadowy way as representations inside our brains. This is the so-called internalist view. Yet, in these discussions, Riccardo Manzotti has frequently insisted that there are no representations in our brains. Rather, he has suggested that experience and object are one. But in that case, what do we mean by an object?

—Tim Parks

Tim Parks: Richard Feynman, the winner of the Nobel Prize in Physics in 1965, insisted that "Everything is made of atoms … and acts according the laws of physics." Any object or animal or person is made up of atoms. Thus if we see not atoms but an apple, a star, or a cloud, this is simply our subjective representation of a reality too intricate for us to apprehend. It seems hard to disagree with this.

Riccardo Manzotti: Not at all. This argument was refuted by Democritus as early as the fourth century BCE in a dialogue between the Intellect and the Senses. Let me quote:

Intellect to Senses: *Ostensibly there is color, ostensibly sweetness, ostensibly bitterness, actually only atoms and the void.*

Senses to Intellect: *Poor intellect, do you hope to defeat us while from us you borrow your evidence? Your victory is your defeat.*

The view that only the smallest constituents, atoms, are "real" is called smallism in science, or nihilism in philosophy, and it clashes with everyday experience and common sense in the most blatant way. As Democritus suggests, it's self-defeating because it is conducted only with the aid of the senses, which it claims have no reality. The world we live in is a world of objects. Apples exist, too!

Parks: But an apple is made of atoms.

Manzotti: To be made of something is not the same as to be identical with it. "To be made of" means that if the atoms were not there, the apple would not be there, either. That is something we all agree on. But the apple is *something more* than the atoms it is made of. The apple—or the car, or the pinstripe suit—exists relative to a human being's body.

Parks: You spoke of objects existing relative to others in an earlier conversation, but I'm not sure I have really understood. Why relative to a human being's *body* and why not simply to a human being? Why not relative to a horse, or a maggot, or a tree?

Manzotti: This is a crucial point, so let's take it slowly. What is required for an atom to be an atom?

Parks: I've no idea. A nucleus. Electrons. I'm no expert on physics.

Manzotti: I mean for an atom to behave as an atom. To function as an atom. To do what atoms do.

Parks: Well, I suppose it needs another atom to combine with. But couldn't it exist without behaving, or functioning, or doing? In blissful isolation?

Manzotti: Nobody has ever found any atom, or any object for that matter, that was not in a direct cause-effect relationship with some other object, influencing it in some way. To be in such a relationship, something science can measure or experiment on, is what it means to exist. The question of whether an atom could exist alone, in the absence of everything else, is empirically unverifiable, and thus scientifically meaningless. Meantime, though, you are right. An atom, to be an atom, requires nothing more than the presence of another atom with which it can combine. In relation, they are atoms.

Parks: But for some larger object, an atom would not be enough.

Manzotti: Right. A handle, to be a handle, requires a hand. A key, to be a key, requires a lock. A face requires a fusiform gyrus that can distinguish a face from mere skin and hair. For a work of art to be a work of art—Michelangelo's *David*, Beethoven's *Waldstein Sonata*, Fitzgerald's *The Great Gatsby*—requires a human being who can see, hear, read. We assume atoms are more fundamental than keys or handles or works of art because they require less work, less stuff, to be what they are. But actually, they are no different from more complex macroscopic objects. Everything is what it is because of its relation to another object.

Parks: And you consider a human being an object?

Manzotti: I deliberately specified a human *body* rather than a human being because the latter is often taken as a synonym for a self, which is a vague notion and something many people may think of as immaterial. A human body, on the contrary, is entirely concrete. It is another object in the mix—extremely complex, of course, but entirely physical. Works of art need human bodies to exist. If there were no one to read a novel, it wouldn't be a novel.

Parks: And you would consider a novel an object on the same level as a stone or a beetle?

Manzotti: I am defining an object as something that comes into being by virtue of its relation to another object. In that sense, yes, we can put stones, beetles, and novels on the same level.

Parks: So, just as a novel isn't a novel when nobody reads it, a stone is not a stone when nobody is around to see it or kick it.

Manzotti: You are hoping to accuse me of Bishop Berkeley's idealism. Obviously, when I am not there to see or touch the stone, the agglomeration of atoms is there in relation to the ground, the air, or another body, perhaps. But it is not the stone I saw.

Parks: It's a different object.

Manzotti: Yes.

Parks: With the same atoms?

Manzotti: Yes.

Parks: But don't we then get back to the formula that reality, minus human beings—*objective reality*, that is—is simply atoms?

Manzotti: Why would you want to subtract the human beings? We are objects, too, and we make things happen by our relations with other objects. As we said, the conditions for being an atom are simple and ubiquitous. The conditions for being, say, a face are far more complex. But both are real, and both are objects. The reason why people have trouble with these reflections is that we have all been educated to believe that things are what they are *in themselves, absolutely.* If you take that line, then inevitably you begin to think that only the tiniest universal particle can be entirely real, entirely sufficient to itself: the atom. But as we've said, even the atom doesn't exist alone. And as soon as you accept the widely recognized scientific fact that an object exists relative to other objects, then it's clear that there are not only atoms, but also apples, cars, and stars. They are *all* viable objects.

Parks: So how does this change the way we think about consciousness?

Manzotti: We live in a time when scientists seem to like nothing better than to expose our everyday view of reality as delusional. They say, "You see the color red, but in fact, out there are only atoms; there are no colors. You hear music, but out there, there are no sounds," etc. This gives them the authority to describe an entirely different reality, in which deciding between chocolate or strawberry ice cream, say, is nothing more than a matter of warring cohorts of neurons transferring their electrical charges and chemical processes this way and that, while outside your brain there is only a flavorless world of atomic particles. It's a vision that denies not only our existence—as people choosing between ice cream flavors—but also the existence of the things we experience: the banana sundae, a new car, paintings, planets, smells, seas. All these macroscopic objects cease to be real. They are all merely subjective. Merely the product of your brain.

Parks: But what if that's the truth of the matter?

Manzotti: It is not the truth. It is a profound misunderstanding. The notion that objects exist relative to each other, brought into existence by each other, does not clash with any scientific finding or demonstrated result. Only with smallism, which, again, is an idea, a theory, not a scientific finding. There are atoms, but there are also macroscopic objects, and the key to understanding why both categories exist and are equally real is that they exist relative to different things.

Parks: Can you give me an example of when the same chunk of physical reality is different depending on the object it is in relation to?

Manzotti: This is not hard. My answer would encompass virtually all the objects that make up our lives. The apple is round and red, but pick it up and close your eyes and it is a smooth, hard thing of a certain weight. A different object. Measure the velocity of a car relative to the road, and it is one number. Measure it relative to a second car, and you have another number. Measure the car's speed relative to a bird, the moon, a passing plane, a distant satellite, and it is different in each case. Velocity is a physical property, yet this property changes depending on the object our car is in relation to. If we have different properties, we have a different object. Because when you see that existence is relative, you also realize that it is *multiple*; the same stuff, which will always be an aggregate of atoms, can simultaneously be different things depending on what object they are relative to.

Parks: But all you're doing is describing subjective experience!

Manzotti: The word *subjective* suggests that a person is somehow inventing what he or she is experiencing, and could perhaps invent it differently. But when I see a red traffic light, I can't *choose* to see some other color. The nature of my eyes, my photoreceptors, and my visual cortex is such that when they encounter this phenomenon, it is red. And the red is out there in the street, not in my brain. The color is not a subjective experience, but a relative object. And my experience is the object that, in relation to my body, is that color.

Parks: Sorry: at the traffic light it's true that almost everyone will see red, but a colorblind person won't. Surely, he or she is simply seeing it "wrong." Subjective.

Manzotti: Why wrong? Because in a minority? The body of a colorblind person is no less real than the body of a person with normal color sight, so the objects that exist relative to the color-blind person are no less real than those that exist relative to any human body. But they are different.

Parks: What if we move away from color to something that can be impartially measured by a scientific instrument? For exam-ple, a thermometer tells me it's seventy-two degrees Fahrenheit. I should feel fine, but in fact, I *feel* cold. At the same time, you're feeling warm. These experiences, which I assume you would call objects, must be subjective since we know that the temperature is seventy-two degrees.

Manzotti: One of the comedies of modern thinking is that we treat objects that exist relative to the tools of scientists as more real, more correct, somehow, than other objects. In the case you mention, there are four objects: the air, and three others in rela-tion to it—your body, the thermometer, and my body. The ther-mometer meets the air and says seventy-two degrees. A digit. Your body meets the air and registers cold. My body meets the air and registers warm. All three "measurements" are valid and real. Seventy-two degrees, cold, and warm. You can't choose to be warm because someone tells you a column of mercury is registering the number seventy-two. There is nothing absolute about the temperature.

Parks: Your point, then, is that we have fallen into the habit of calling objective and real something that exists *in relation to* a scientific instrument. And we call it subjective and possibly hallucinatory if it is in relation to an individual body that experiences it differently from others. You're claiming that this is a form of cultural discrimination, not a scientifically useful distinction.

Manzotti: Exactly. And notice, in contrast, how democratic this notion of relative existence is: the average human body, the blind person, the deaf person, the dog, the scientist's instrument are all equal conditions for bringing an object into existence. As in physics, there are no frames that are *more true* than other frames, only frames that make it easier for us to compare certain situations. The thermometer simply makes it easy to compare seventy-two degrees with one hundred and ten.

Parks: Has anyone else ever put forward this view of existence as relative?

Manzotti: In one of Plato's more difficult dialogues, a rather mysterious, unnamed philosopher referred to as "the visitor" or "the stranger" argues that existence is a form of action, of doing things. Since the visitor was from a place called Elea (corresponding to the present-day village of Velia, in Italy, not far from

Pompeii), this notion came to be called the Eleatic Principle. It has been restated many times over the centuries, using slightly different terms, by philosophers as diverse as Samuel Alexander, Jaegwon Kim, Trenton Merricks, and Peter van Inwagen. They all connected existence with cause-effect relations between objects.

Parks: Can you give us the mysterious stranger's exact words?

Manzotti: The reference is *The Sophist*, 247e: "I say, then, that a thing genuinely is if it has some capacity, of whatever sort, either to act on another thing, of whatever nature, or to be acted on, even to the slightest degree and by the most trivial of things, and even if it is just the once. That is, what marks off the things that are as being, I propose, is nothing other than *capacity*."

Parks: It seems hard to get from this fragment to the notion that consciousness is outside our heads.

Manzotti: Perhaps it's time to ditch the word "consciousness" and simply talk about experience. You're in the kitchen looking at an apple on the table. It exists *qua* apple in relation to your body. When you are out of the room, it occupies a space in air; it weighs on the table; it reflects light. But there is no apple. Because what is actually there depends on its interaction with things. Your body is such a thing and when your body is there, an apple is there, too. Not an apple reproduced like a photo in your head. An apple there on the table, in relation with your body.

Parks: So, anything the body experiences as an effect—which is to say, anything it experiences—is an object?

Manzotti: The body does not "experience an effect." The experience *is* the apple, which is the cause of an effect. My body allows this agglomeration of atoms to become the cause of an effect *in my body.* Such a cause is the relative object, the apple, and it is my experience, too. Check out the figure below. How many crosses are there?

u	*u*	*n*	*u*	*u*
u	*u*	*n*	*u*	*u*
n	*n*	*n*	*n*	*n*
u	*u*	*n*	*u*	*u*
u	*u*	*n*	*u*	*u*

Parks: I see only one. There's a gray cross, and there's a grid with characters.

Manzotti: Look more carefully: you'll see that the central horizontal and vertical axes of the grid are all *n*s while the other cells are all *u*s. Look hard and you'll begin to see a cross of *n*s. The cross on the left is an object that exists relative to a simple state of things: most animal eyes that can distinguish between gradations of gray would perceive it. The cross on the right exists relative to a much more complex object, a human body.

Parks: But do you claim that all experiences, thoughts, memories, dreams—all the things we refer to as mental—are also relative objects?

Manzotti: I do, indeed.

Parks: Even if thoughts are obviously not made of atoms.

Manzotti: Let's leave that interesting question to our next conversation.

13. THE PIZZA THOUGHT EXPERIMENT

"I think therefore I am." Descartes declared thinking the ultimate reality, the only way to be sure of one's existence. He located that thinking in the mind and believed it was immaterial, made of spirit, communicating with the physical body through the pineal gland at the top of the spine. Times have changed, and scientists now look for explanations of the experience of thinking in the billions of neurons in the brain with their trillions of electrical connections and chemical processes. Yet the location of thought remains firmly in the head. Challenging this, Riccardo Manzotti has suggested that our experience actually lies outside our bodies, one with the objects of our perception. He has extended this to dreams and hallucinations, treating them as cases of delayed and muddled perception. But surely thinking and, in particular, the internal monologue that most of us live with and call the self can't be imagined as taking place anywhere but in the narrow confines of the skull...

—Tim Parks

Tim Parks: When one thinks of the richness of conscious life, that constant overlapping of thought and perception, the words that go back and forth in our minds, as we walk down the street, or close our eyes to sleep, it is very hard to see how it could be conceived as having a reality outside our bodies.

Riccardo Manzotti: We need to establish what we mean by a thought. Have you ever seen a thought, for example? Has any scientific instrument ever identified one? Can you build a thought detector? More particularly, do you really experience a thought, any thought, as separate and different from the object of that thought?

Parks: You obviously want me to answer no to all those questions. But I'm not sure about the last. I do have the impression that I experience thoughts and that what I experience is not identical with the thing I am thinking about...

Manzotti: In that case, the thought is very likely to be identical with the language in which it is framed. Basically, we have two options: either we think things directly, or we think words and sentences about things. I suggest we put language on hold for a moment and concentrate on thinking things and situations directly. When I think of, say, the Coliseum, what is my thought of the Coliseum if not the Coliseum itself?

Parks: You are going to try to include thoughts in the category of prolonged, deferred, or reshuffled perception, as you did with dreams and hallucinations. Some previous contact with the Coliseum, or with a photo or film of the Coliseum, or with something written about the Coliseum, continues to act on my body and I think the Coliseum.

Manzotti: Right. There is no such a thing as a thought that lies between your body and the Coliseum, or the photo you saw of it, or the article you read about it, no need for some sort of immaterial thinking magic to connect your actions with the external world. Simply, there is your body and there is the external thing your body has been in contact with. Of course, there's also a lot of neural machinery that allows the world to produce effects through the body, but the experiences we call thoughts are no more, no less, than the external object as it affects the body.

Parks: It seems to me that you are constantly seeking to reduce what most people call mental life to a direct experience of the material world, yet, in order to do so, you yourself come out with complex and provocative formulations that can't easily be brought back to any original contact with the world.

Manzotti: Let's try an experiment. Think of something, right now. Anything.

Parks: Okay.

Manzotti: Now describe the thought you've come up with.

Parks: Eggplant and parmesan pizza.

Manzotti: That's not a thought. That's what you thought *about*! I didn't ask you to describe the object of your thought. I told you to think of something, then describe the thought, not the something.

Parks: Hmm. Pleasing. Mouth-watering. Steamy.

Manzotti: These are all adjectives that apply to the pizza or to the body's reactions to the pizza. What you like to think of as the thought is simply the object.

Parks: But when I move away from an object, my thinking allows me to accomplish all kinds of things I couldn't do without thinking—not least, engaging in this conversation.

Manzotti: Do you think when you play chess?

Parks: I hope so. Though I'm a poor player.

Manzotti: You think things like, "If I move my knight to capture his bishop, I risk exposing my king to a queenside attack from the rook presently lurking behind a pawn shield"?

Parks: Stuff like that.

Manzotti: And you call that thinking?

Parks: What else could you call it?

Manzotti: Over the last fifty years, computers have been developed that play chess far better than you do, not to mention solving mathematical problems, driving cars, and so on—achievements you associate with thinking. But do computers have thoughts? Of course not. Has anyone ever detected a thought inside a computer? No. There are no courses or exams about thoughts when you take a degree in IT or AI. What we call thinking is a form of action, a way our body organizes our behavior in response to those external causes that our so-called thoughts are about.

Parks: But the computer plays without any experience of playing. It doesn't say to itself, "Huh, now he's got me in a corner." It doesn't rejoice when it wins, or curse when it loses. It doesn't even know it's playing chess. It just acts out a series of instructions.

Manzotti: I didn't say that computers were conscious. However, computers show that cognitive skills do not require thoughts. You do not need immaterial thoughts to choose in a chess game which move is best. It is a physical chain of causes and effects that starts with external objects and ends with actions.

Parks: It seems now you're distinguishing between a thought, which you say is simply at one with whatever object the thought is supposedly about, and then this activity—most people might call it "mental activity"—that allows us to hop from one thought (one object, as you see it) to the next, connecting them together. I need some clarity here.

Manzotti: The notion of "thoughts" has much the same function as the notion of "luminiferous ether" in the nineteenth century. Scientists couldn't understand how light could travel, so they invented this mysterious medium that somehow propagated light. But just as light does not need ether to travel, so objects do not need thoughts to have causal effects on our bodies. When I say, "I think of x," it is simply a manner of speaking to explain that x is exerting effects through my body.

Parks: But thoughts are different from one's immediate perceptions of the world. When I think of, say, my daughter, it is quite different from actually seeing her, or even dreaming of her.

Manzotti: Absolutely. When you think of her, the object that is identical with your experience is different from the object you see when she is physically present. Just as reaching a hand to touch something inside a drawer and then looking in the drawer leads to different objects of perception—the object, remember, being relative to the perceptive faculties, not absolute. So you could think of thought as another form of perception that allows a different object to affect your body. The way touch, sight, and hearing allow different objects to affect us.

Parks: As I recall, Buddhists believe we have six senses, the mind being the sixth, and thoughts being the things the mind perceives.

Manzotti: I'm afraid I know very little of Buddhist ideas, but yes, you could say we're talking about another group of objects of perception. For instance, when I see an apple, it's easy to pinpoint the external object that is affecting my body, just as when I hear or touch something. But what about those situations where what is acting on me, my experience, is a combination of various objects, separate in space and even time?

Parks: For example?

Manzotti: A constellation of stars is perceived as a single shape, despite the distances between the separate objects and the fact that they are only in that apparent spatial relation thanks to the different times traveled to reach us. Each star, though, is there in time and space. More mundanely, objects behind and reflected in a shop window act together to make a whole object of perception. Or again, a melody spread across time is combined into a unity.

So why should we put a limit to this process? What if the causes of one's actions, one's behavior, were an extended collection of external causes, deriving from external objects, or even the relations between them? Such is the case when I react not because there are two specific things, but because the relation between those things acts itself as a new object. You're on a long walk, you're only wearing a T-shirt, a raincloud appears in the sky. The relationship between these things presents itself as a new object: fear of getting cold and wet. Of course, it's not easy to ascribe physical properties to a relation like this, but with the right physical system, that relation can nevertheless exert an effect.

Parks: The right physical system being?

Manzotti: The human body with its extraordinary brain. Beyond the so-called early sensory cortexes, there are many cortical areas whose functions remain obscure. Very likely, they are allowing the external world to come together in novel ways that correspond to complex external objects. You could think of the brain as a huge eye that, by reorganizing its neural structure, is able to perceive all kinds of complex objects across space and time.

Parks: And you're saying all this connecting up is what we call thinking.

Manzotti: If hard-pressed, I'd say the external objects were the thoughts, and their churning over the thinking. But I'd prefer not to use these words because I'm trying to describe what happens without bringing in fictive mental entities. Let's just say that the human cortex with its billions of neurons has trillions of "gates," each of which is causally continuous with the external world. These gates keep reorganizing their pathways so that the objects that affect our body and behavior become those that best match our needs.

Learning is a process of adding and connecting such objects— objects that are really out there, not in the mind. The internal activity of the brain allows the objects that are us—and we are not our brains, remember, we are our experience—to be more complex than those singled out by the immediate senses. The body intercepts separate notes struck on a piano, the brain allows the sonata to affect us as a single object, which it is, relative to our brain. In general, objects that would otherwise be separate—something visual, something tactile, something auditory—become a whole and affect the body as a single object, a crunchy candy bar, for example. Because, relative to our body, they are a single object.

Parks: And presumably, this constant mixing and remixing of the world has a goal.

Manzotti: Of course. To behave in a more successful or appropriate way. This is entirely in line with an evolutionary view of the brain.

Parks: So what we call thinking is a practical activity, a churning of causal paths, a reshuffling of the world, the making of causal connections between the world and our actions. While direct perception—seeing, hearing, touching—is unavoidable. It simply is our experience.

Manzotti: In the past, philosophers usually assumed thinking to be a superior cognitive skill capable of penetrating the essence of reality. So thinking was associated with truth, while individual experience was dubbed subjective and downgraded to mere appearance. In fact, the opposite is the case. It is our direct individual experience that is unerringly true; being one with the external world, it cannot be wrong.

What we call thinking, on the other hand, is the way in which the world and our behavior are combined and, as such, it can all too easily be wrong. For example, I may *think* that a mushroom I've picked in the woods is edible, but when I eat it, I'm poisoned and die. This wasn't a mistake in my visual or tactile experience of the mushroom, which simply was what it had to be, but rather in the pairing between what I saw and what I did.

Parks: I can't help feeling that we're now talking about thinking as going on in the head. As if there were the experience of the mushroom outside and the thinking inside.

Manzotti: Of course, there is always a great deal going on in the brain, but thoughts remain external objects, experiences, which exist in relation to our body, of which the brain is a part. Thoughts are just more complex objects created by combining a number of objects together. In this case, we have the mushroom, which is a rather simple, proximate object; the appetite of the body, which is a property of the body, itself an object; and past occurrences of eating mushrooms, again events obviously involving objects and capable of still acting on the body. These come together in a single object that we dub a desire to eat the said, alas poisonous, mushroom.

Parks: What about creative thinking?

Manzotti: Most of what happens in our brain, the connections it's constantly bringing about, is unconscious. So creative people do most of their thinking while they do other stuff, or even while they sleep. There is abundant anecdotal evidence of the brain reaching solutions without our being aware of it. "A thought comes when 'it' wants to," says Nietzsche, "and not when 'I' wish." So thinking is a process of physical reconfiguration of causal paths, a complex form of perception. To use the word "mental" in opposition to physical, in order to give some special status to this process, only confuses matters.

Parks: So there's no great merit in being creative.

Manzotti: Creativity is highly prized because it allows all kinds of novelty to occur, but it doesn't necessarily require either conscious experience or even a mind. After all, natural selection has created millions of incredibly complex structures. Yet there is no universal mind at work, just an endless permutation of DNA bases singled out through practical operations. Thinking is the same.

Parks: Very often, though, we are indeed aware of thinking, aware of a voice chattering in our head. Particularly when I write, I'm aware of moments of breakthrough, moments when I've arrived at some new synthesis, perhaps. It is all very conscious. Sometimes, painfully so. How does this fit in?

Manzotti: "Voice," you say. We're back to language. It's obvious that there are thoughts that can only come in words. The formulation "thoughts are imaginary entities," for example, is an entirely verbal construct. It even includes the notion of something that, by definition, cannot be an object in the world: an imaginary entity.

Parks: So? How can this thought be the result of an object in the world? Unless you want to say that language or words are the object. But in that case, where are these words? Since you love to say that everything is physical and hence everything is located somewhere, you will have to tell us where these words are. And where, indeed, if not in my head?

Manzotti: Ha. I will tell you all that, and more, but in our next and penultimate conversation. Language is too big a subject to tack on in a last paragraph. Let me leave you, though, with a quote from E.M. Forster to mull over: "How can I tell what I think till I see what I say?" If we understand what he meant by that, I believe we have the beginnings of an answer to your question.

14. CONSCIOUSNESS: WHERE ARE WORDS?

Words, words, words. With the advent of the stream of consciousness in twentieth-century literature, it has come to seem that the self is very much a thing made of words, a verbal construction forever narrating itself and reconstituting itself in language. In line with the dominant, internalist view of consciousness, it is assumed that this all takes place in the brain—specifically, two parts known as Wernicke's area and Broca's area in the left hemisphere. So, direct perception of sights and sounds in the world outside the body are very quickly ordered and colored by language inside our heads. "Once a thing is conceived in the mind," wrote the poet Horace in the first century BC, "the words to express it soon present themselves." And we call this thinking. All our experience can be reshuffled, interconnected, dissected, evoked, or willfully altered in language, and these thoughts are then stored in our brains.

—Tim Parks

Tim Parks: Riccardo, you have been presenting quite a different theory of consciousness, which suggests that there is nothing "stored" in the brain. No images, no sounds, and, I assume, no words. As you see it, the brain is a powerful enabler that allows experience to happen, but this experience is actually one with the objects we see, hear, smell, and touch, outside our bodies, or again, one with the body itself in those cases where it is the body that is experienced. And we are this experience, not something separate from it. You have explained dreams, hallucinations, and non-verbal thoughts as more complex and delayed forms of perception, always insisting that each experience is identical to something outside the brain, something whose existence is relative to our body. But how can that possibly be the case for language? When I am thinking silently to myself, where can the language possibly be, if not inside my head?

Riccardo Manzotti: Imagine you're lying in bed planning to furnish a house you'll soon be moving to in a distant town. What do you do? You start thinking about different items of furniture to figure out if they will go together in the space there. This would normally be explained by saying that you are imagining mental objects—representations in your head—and arranging them together in a mental space, another mental representation, again in your head. But in our previous conversation, we reached the conclusion that there are no "mental" objects—no thoughts, that is—separate from real objects. Simply, there is no need to introduce this new entity, thought, between body and object.

When we say we are thinking, what we are actually doing is rearranging causal relations with past events, objects that we have encountered before, to see what happens when we combine them.

We don't need a mental sofa to put next to a mental armchair. We allow the sofas and armchairs encountered in our past to exert an effect in the present, in various combinations. Like a controlled dream.

Parks: But we were talking about words, Riccardo, not sofas and armchairs! Last time we talked about thinking things directly; this time we're considering thinking *in language*, which is surely different.

Manzotti: Not at all. Words are really not so different from sofas and armchairs. They are external objects that do things in the world and, like other objects, they produce effects in our brains and thus, eventually, through us, in the world. The only real difference is that, when it comes to what we call thinking, words are an awful lot easier to juggle around and rearrange than bits of furniture.

Parks: External objects you say. But what kind of object is a word? Is it a sound? Is it a visual sign? What about when it is neither spoken nor written, but simply thought in the head?

Manzotti: Exactly as with the furniture, what we have is a rearrangement of our causal relations with past events—in this case, words initially heard in the external world. If we take things slowly and simply observe what happens as we learn to speak and think, you'll see that, once again, there's no need to posit the entity you call "a thought in the head."

Parks: Go on, then.

Manzotti: A mother repeats words to a baby. That's how it starts for most of us. So, we have sounds, which are physical objects. The sounds are repeated, frequently, insistently, in reference to things, actions, emotions, to the point where they become labels that are perceived together with facts in the world. As soon as you have the fact, the sound/word is present; as soon as you hear the sound/word, the fact is present. And when you make the right sound, the food arrives.

Parks: I suppose you could say the same thing happens with certain movements that a parent teaches a child. Pointing, waving, clapping, or even using tools: spoons, forks.

Manzotti: Indeed. But it's important that the sounds/words don't just come singly; they come in patterns, building up more and more complex objects, phrases, sentences. And many of the sounds don't relate to an object/experience, but to the other sounds in the pattern. I mean words like, "and," "but," "as," "because." Then the patterns can be rearranged. Hear the beginning of the pattern and the rest follows automatically.

Parks: Surely not altogether automatically, or we would simply be forever repeating the same things?

Manzotti: True. Let's say there are powerful automatisms, rules if you like, that govern this constant rearrangement of words we experience. The further you proceed in a sentence, the more your options close down. However, we're not trying to offer a linguist's description of language function here, simply to establish how and where language is experienced. It is a patterning of real, physical, external objects, and these are one with our experience. They *are* our experience.

Parks: But some words refer to imaginary entities, or abstract entities. Are they external objects?

Manzotti: All sounds/words are physical objects in the same way. But you're right, once we get used to the idea that a sound refers to an object, we can start inventing sounds/words for imaginary or notional objects—angels, dark matter, whatever—and then use other sound/words to put them in relation to the experience/objects we are familiar with.

Parks: But how did we arrive at an imaginary or notional object, something that doesn't perhaps exist, if you're saying there is nothing to experience but real external objects?

Manzotti: What is an angel but a juggling of past experience: beautiful body, plus wings, as in a dream? What is dark matter if not a piece needed to complete a puzzle, a theory, made up of endless complex objects in the world? Sometimes, the imaginary object is a reshuffling of real objects and thus it is real in its own way; sometimes, it is nothing. Both these very different notional entities can be traced back to direct experience.

Parks: In other words, you're saying that we don't so much think in a language—English, Italian, whatever—but constantly rearrange our experiences, our world, of which the language we speak is a part. It's world-ish!

Manzotti: Yes. Think of people who are born stone-deaf. Are they going to be thinking in sound/words? If they've never heard words, how can they be hearing them in their thinking? They can't. In fact, if you ask such people about this—in writing, of course—they tell you that they think in signs. They've learned their language visually, so they think words visually. Those are the external objects they are dealing with. Interestingly, some people born deaf say they think in images with captions.

Parks: Which brings us to the written word.

Manzotti: Right. At a certain point, the child who has learned to speak is invited to look at signs on paper or on screen, and relate them to the sound/words he or she knows. This introduces yet another object into the mix: the written word. Written words now prompt sounds and thus the child's relations with object experiences. Words also facilitate the rearrangement of these experiences in all kinds of extended forms. We go back and forth so frequently between these different kinds of objects that we have put in strict relation to each other—sound, sign, referent— that as the world acts on our brains in what we call experience, so the sounds that we relate to each object/experience are also active in our brains. And we call this silent thinking.

Parks: To sum up, words and language amount to a kind of behavior—like learning a series of movements, or a dance, with infinite variations.

Manzotti: Correct, and just as you can't find a pirouette stored in a dancer's brain, so you wouldn't have found "To be or not to be" stored in Shakespeare's. At the end of our last conversation, I quoted E.M. Forster's remark, "How can I tell what I think till I see what I say?" This is exactly right. Language is a performance. We launch into it, we have a direction and we set off. But until we have said or written the thing, until we've performed it, we don't know how it will turn out.

Parks: So, nothing is stored in the head.

Manzotti: All the objects we encounter, the objects we call experience, continue to be active in our bodies and brains, continue to be our experience. It is the nature of our fantastically complex brains that they allow these encounters to go on, and to go on going on. The encounters are not "stored" and are certainly not static. They are continuing to happen. They are us.

Parks: But there must be some order. Isn't this a recipe for chaos?

Manzotti: Words are one way we try to impose a rather crude order on a fantastically complex, nuanced experience. Language makes categories—for example, in the area of colors, where we name only a small number of the vast range of colors we see. In 1858, the British statesman William Gladstone, commenting on Homer's references to color in *The Iliad* and *The Odyssey*, suggested that our experience was limited and determined by linguistic conventions. But in 1898, the ethnologist W.H.R. Rivers's study of the inhabitants of Murray Island (between New Guinea and Australia), people who had no names for colors as basic as green and blue, conclusively demonstrated that human beings see all shades of colors even when they don't have words for them. In 1969, the anthropologists Brent Berlin and Paul Kay established that color names do not change the colors one sees, and later studies have confirmed this.

Parks: So, by segmenting infinitely nuanced experience into a limited number of words, language allows us to impose a little order, though presumably at the risk of leaving a lot of experience out of the system.

Manzotti: It's like a net we've made and throw over the world. It catches some fish and lets others slip through the links. But it doesn't create fish, or stop them from existing. And we're always tampering with our net to have it capture new fish, or exclude certain fish we want to pretend don't exist. Sometimes, the net captures fragments of the net itself (for the net itself is an object), and it gets tangled and snarled. In short, it's as if we'd built a world with Lego bricks that corresponds very crudely to the world as a whole, and we keep adjusting it to fine-tune the correspondence.

Parks: And just as there is no inner "cinema screen" in the brain on which visual images are seen, so there is no inner auditorium in which the voice resounds.

Manzotti: It is all echo and learned behavior. Our words are the words of our mother tongue, not other languages (or not until we've been exposed to them for a long time). There is some interesting overlap here with the work of Julian Jaynes, a Princeton-based psychologist who in the late 1970s claimed that consciousness is not concocted inside the brain but is a learned process based on language. As Jaynes saw it, we talk to ourselves and hallucinate our own voices. His hunch was that, in the past, people have often mistaken this hallucinated voice for external sources, thus inventing gods, angels, demons, and the like.

Parks: Whereas you, I presume, would see all supposedly internal voices as simply a collection of past voices still rumbling on thanks to the brain's capacity to allow immediate perception to continue.

Manzotti: Yes. In the 1950s, when the neuroscientist Wilfred Penfield used electrical charges to stimulate the cortex of his patients, they reported hearing the voices of people they knew. Since then, other scientists have consistently obtained similar results. As we said when we discussed dreams, all perception involves both immediate experience and, as it were, buffered experience, perception that remains in the background and comes to the fore, if at all, only when immediate attention to the here-and-now is not required.

Parks: But in that regard, words are rather different from, say, my visual memory of an exciting soccer game, in that I keep performing language pretty much all the time, and when I want. If it's a hallucination, it's an extremely controlled hallucination.

Manzotti: Yes, because we train ourselves assiduously and from the earliest age to give structure to the world in this way. Learning languages, even dead languages like Latin and Greek, has always been considered good for one's education, because it promotes this organization of experience. Society demands language of us and promotes it most determinedly. In those very rare cases where a child is abandoned in nature, yet survives (so-called feral children, or wolf children), without having gone through the language-learning process, they are not, alas, like Kipling's Mowgli, smart, bright kids. Unused to the constant manipulation of experience that language allows, they are cognitively impaired. Language, as Ludwig Wittgenstein insisted, is a game. Constantly playing this game, we learn how to organize those objects we call words and, eventually, with them, the world.

Parks: As always, you have systematically eliminated any entities that might intervene between the world and the body, all representations and so-called mental phenomena. As you see it, all experience is made up of external objects and the rearranging of external objects. But what I want to ask now is the crunch question: Who is doing that rearranging, who is focusing on this part of the landscape or that, who is choosing this word rather than that?

Manzotti: Where and what is the self you mean? Who says I? Who pulls the levers? A soul, a ghost in the machine? We'll tackle this in our next, and last, conversation. But let me leave you with that all-too-famous quote from Ralph Waldo Emerson as he crossed a bare field in winter twilight:

I become a transparent eye-ball; I am nothing; I see all; the currents of the Universal Being circulate through me; I am part or parcel of God.

But instead of the Universal Being, God, perhaps we could just say the world: "I am a part of the world, that part relative to my body."

15. CONSCIOUSNESS AND THE WORLD

We are all used to thinking of ourselves as a single entity that enjoys a certain continuity from birth to death, if not beyond. We confidently say, "Ten years ago, I did this; next week, I will do that." But what exactly is this entity? Traditional Christianity posits a nonmaterial soul housed in a material body. At the beginning of the scientific age, Descartes formalized this concept with an idea that came to be known as dualism: the soul in the skull was spiritual and interfaced with the material world of the body through the pineal gland between brain and spinal cord. Modern materialism jettisoned this idea, yet struggles to replace it; even the most serious philosophy and neuroscience publications often lapse into an implicit dualism, where, for example, the word "mental" is opposed to "material" without our being quite sure what we mean by the two, or where consciousness, and with it the self, are understood as "supervening" on vast quantities of "information" moved back and forth in the brain. It is very much as if the brain were made equivalent with the self, and information afforded some non-material, "mental" existence. "You are your brain," claims the neuroscientist David Eagleman.

—Tim Parks

Tim Parks: Riccardo, in these conversations you have been elaborating a radically externalist theory of consciousness; rather than representations in some inner theater of the brain, experience is understood as united with the world that comes into contact with our bodies, allowing no separation between subject and object. In this view, nothing is stored in the head, nothing is "mental." Simply, we *are* our experience, and it is all out there. But if we are no more, no less, than our experience, what about the self? How can I be someone, if I don't have a self in my head?

Riccardo Manzotti: If you *had* a self, the self would not be you, right? But let me ask you, have you ever seen, or in any way observed, or, even better, pinned down someone's self? Or your own self? Of course not. No one has.

Parks: I observe behavior I'm familiar with in people I know, alongside facial expressions and gestures I've seen a thousand times. I know when a friend is acting himself, or herself, as they say. Similarly, I know my own way of living and thinking, all too well.

Manzotti: That makes more sense. Let's focus on a concrete example. Yesterday evening, I was watching my two boys preparing for a party. One is eleven and the other is eighteen. They were arguing as usual, the younger accusing the older of wearing his pants in a stupid way, sagging off his butt. I must say I rather agreed with him, but no doubt my eldest is aping his friends' way of dressing. It's a fashion. So, was I seeing the boys' "selves"?

Parks: A thing doesn't have to be visible to exist. We know a dark hole is there because of its gravitational pull on the area around it, not because we can see it. Having met your kids a couple of times, I can only say they both radiate very distinct and attractive identities.

Manzotti: From selves to black holes! But yes, there are cases where certain goings-on require us to posit a physical entity we cannot see; in those cases, we make a hypothesis and set to work with all our scientific instruments to pin this entity down. But does this behavior of my children really require me to suppose there's an invisible "self" around that we should start looking for? Aren't there any number of entirely visible causes to look at before positing something invisible?

Parks: Like what? And remember, it's not just the behavior, but the awareness and sense of purposefulness that accompanies it. Or are you just going to say what Daniel Dennett said about "intentional stance"—his term for selfhood, I think—that it is only in the eye of the beholder? A descriptive tool?

Manzotti: Not exactly, no. The notion of an internal self is akin to that of the hypothetical center of mass they use for calculations in mechanics. It's a form of shorthand. It doesn't exist as such. But the cause of the ongoing behavior pattern, as you put it, coupled with awareness, etc.—the *person* even, if you like—certainly does.

Parks: So, what is it, physically, scientifically?

Manzotti: You could say that, at any moment, a person is the collection of all those objects that are presently active thanks to a particular body. Or, again, it is the world that exists, now, relative to a certain body, an immensely complex amalgam of things spread in space and time, but all, at one moment or another, perfectly available to the senses.

Parks: I'm losing you now. How can I possibly think of a person in this way?

Manzotti: Let's get back to the example we began with: my boys arguing over the way one of them wears his pants. We've agreed in previous conversations that since there is nothing in the head but neurons and chemicals, the brain can't constitute our experience; the only thing *identical* with a person's experience is experience itself, which is outside our nervous system. When I see an apple, my experience is the apple I see and I am my experience. This is why objects of whatever kind—cars, songs, other people, pets, food—are so important to us; they constitute our experience.

Parks: And so, your boys?

Manzotti: Giulio, my older son, is eighteen. For years and years, every waking moment, inescapably, his body has been in contact with the world that has become his experience and that continues to produce effects by means of his body and brain.

Parks: That's an expression you keep using; what exactly do you mean by it?

Manzotti: You hear a song yesterday, and you sing it today. You watch an ad for a brand of sunglasses and, six weeks later, you buy them. You remember your mother's voice from twenty years ago. You dance a step your friend taught you in your teens. You...

Parks: Enough, I've got it. These were all originally objects in the world, experiences my body is still experiencing.

Manzotti: Think of the body as a kind of hyperactive proxy that allows objects in the world to manifest themselves, much the way your phone allows a friend's voice to be heard. The cause of the voice is not in the phone, but the phone is necessary for the voice to be there.

Parks: Except the body would have to be a proxy and go on being a proxy for any number of things at the same time.

Manzotti: Right. A whole world speaks through your body and that world is consciousness, the consciousness you think of as yourself.

Parks: Back to Giulio.

Manzotti: Okay. Giulio's multitudinous experiences include his little brother, Emilio, his school friends who wear their jeans sagging ridiculously low, the jeans themselves, of course, and his own body, which is a constant object of his own experience. If we want to use the word self, we could say that rather than some ghostly entity imprisoned inside Giulio, self is a shorthand for all these things acting together through his body. It is his *life*.

Parks: I appreciate your desire to avoid the mystification that lurks behind words like self and soul. But this description seems hopelessly incomplete. If I am nothing other than the myriad objects of my experience, why would I be different from anyone else who has come into contact with the same objects, more or less? Why would we have this powerful sense of distinction between different people, and this awareness of a continuity of behavior in our own lives? How can we have character, or even temperament, in your vision of the world? It seems to me it's this awareness of individuality that leads people to posit the existence of an internal entity, self or soul.

Manzotti: First, each of us has a distinctive body that comes into the world in distinctive circumstances, setting us up for a unique set of experiences—foremost, of course, of the body itself, which is not only the agent of experience, but also the prime object constituting experience: libido, appetite, pain, pleasure, illness, and so on—then, of all the other objects relative to that particular body, including parents, aunts, uncles, brothers, sisters, physical and social environment, school, etc. Then, day by day, we are constantly performing, repeating, and subtly altering areas of our experience—language, learned behavior, and so on—while all our past experience conditions each new experience.

As a result, the apple I grab is different from the apple you grab—even if, from an eating perspective, there is just one apple—because my apple is relative to my body and experience, and yours is so to yours. If you were blind, your apple would lack color. The poem you read is relative to your life experience; the poem I read is relative to mine. Maybe my moon (or the moon that is me, at that moment) is pretty much the same as your moon, visually, but who knows what past moons and experiences make the two quite different? If we look at life this way, it is actually far easier to explain distinctions between people than if we posit some soul or self already there at birth.

Parks: So, both Giulio and Emilio, thanks to their bodies and brains, bring into existence a world of relative objects, experience, and habitual patterns of juggling those objects and having them interact with new objects. And that is what you call the self? Have I got it?

Manzotti: Right. Though I prefer the word "person," or just "life." In any event, this amalgam of experience is relative to the body, but not *inside* it and not *identical* with it. Above all, it's not "mental," and it's not a fixed entity. It has a certain continuity, or inertia, but it is also open to change.

Parks: Help me with the "not mental" side. Let's say that as the two boys argue about these sagging pants, Giulio is thinking, "Emilio is only saying this to please Dad." And you are thinking, "How can I show my agreement with Emilio without seeming a boring old conservative?" And Emilio is thinking, "I'm going to be really embarrassed turning up at this party with my brother wearing his pants down to the floor." And at the same time, all three of you are looking at the pants and one another in the mirror of the wardrobe. Don't we have an awful lot of mental stuff going on here? Then, at a certain point, Giulio turns to Emilio and says,

"You're just jealous!" What was it that decided him to do that? What orchestrates this amalgam of experience as you've called it? Don't we have to posit some unifying entity to explain this?

Manzotti: Nice drama. Let's observe two things: first, communication; then, the action that you say demands a "mental self." The common notion of communication supposes that people exchange packets of symbols—words, gestures—that their inner selves then interpret by associating each symbol with the correct meaning. This view is a myth. Curious as it may seem, our current models of communication are derived from the work of mathematicians like Claude Shannon and Alan Turing who, during World War II, were involved in such things as deciphering coded messages between enemy submarines and their naval commanders. People do not communicate like navy submarines.

Parks: You're telling me we don't exchange messages?

Manzotti: Of course, we send things back and forth. But communication is not an exchange of messages to be decoded. Communication is sharing. It is the activity through which different people—by which I don't mean people's bodies, remember, but their experiences—become composed of the same stuff.

Parks: So, as you all look in the mirror, contemplating one another and Giulio's sagging pants, you are communicating.

Manzotti: Absolutely. We are our experience and those sagging pants are shared between us. But people do this all the time when they point things out to each other. Look at this, look at that. A seaplane landing on a lake. A strange orange spider. Or listen to this, a new song by a favorite band. Touch this, smell that. This is communication. The airplane or spider, relative to our bodies, becomes our shared experience. Likewise the TV, the radio, the new *Star Wars* film.

Parks: But we could communicate these things in words, speaking or writing.

Manzotti: If someone were blind from birth, no amount of words, or bundles of bits in a computer processor, could convey the orange of the orange spider. Communication is shared experience, being literally made of the same world relative to our bodies. With words, we share the experience of words and their appeal to previous experience. "Orange spider" in words will activate a previous experience of orange, but not maybe the exact hue we share when we look together.

Parks: But is it communication even when you disagree? About the pants, I mean. Doesn't that suggest the experience isn't entirely shared, in the sense of being the same for each?

Manzotti: The new object, as I said, is *relative* to me, my life, my experience. Maybe I had a bad experience with a spider as a child, so my body doesn't react to the orange spider as yours did. Maybe I have a long experience of jeans worn in a certain way and don't have friends I'm eager to impress who wear theirs sagging to their knees. Still, this is communication. We share the same world and declare our difference by our different, sometimes contrary, responses. That response can then modify the other's experience. I see the sagging pants and laugh, and maybe Giulio now experiences the pants differently. I jump seeing the spider, and maybe you now experience it differently. A critic pours scorn on a book you like, and you begin to see it through his eyes and have your doubts. The book you now see is a new book relative to you. So, as well as sharing, communication is change, experience is accumulated, shifted. Which is why, as well as continuity of identity, we also have discontinuity.

Parks: What about decisions? I decide to focus my eyes on this, rather than that. To listen to my friends, rather than to my father. To suddenly say to my brother, "You're just jealous." How does that fit in?

Manzotti: We can't stop acting and being, moment by moment, and the amalgam of experience that we are reacts constantly to the new experience of the moment. Emilio laughs at the sagging pants; Giulio reacts with the old card of accusing the younger boy of jealousy. There are any number of stories going on at once, worlds manifesting themselves through the proxy our bodies offer.

Parks: It seems you always see action as reaction, you always explain decisions in terms of the influence of the relative objects, experiences, constituted by our body's meeting with the world. But doesn't this seriously diminish us? Who are we, finally? Who am I? Who are you? Just a bundle of reactions?

Manzotti: We have an interesting linguistic trap here, one created by centuries of human self-regard. By using a different pronoun to enquire about the identity of people rather than of things—who, instead of what—we introduce an imaginary metaphysical difference. Why not ask: "What are we? What am I?"

Parks: You mean, essentially, that we are objects, and objects "take place," rather than act.

Manzotti: We are part of the physical world, hence objects. What else could we be—immaterial souls?

As for identity, we are what we are because we are identical with a portion of the world that has come together over the years in a certain way. The traditional separation of subject and object that underpins all standard thinking on consciousness and identity lies at the heart of our troubles as individuals and as a society. Convinced that we are separate from the world, we feel we have been expelled from the Garden of Eden, and we yearn to return, maybe after death. But however useful the subject-object divide may be for all kinds of practical matters, it is plain wrong.

Parks: So, a subject is never in relation with anything but what that subject is?

Manzotti: I would say something like, "Thou shall have no other relations but identity," and, no, I'm not starting a new religion; it's just the most elementary fact of physical reality. In nature, everything is what it is and only what it is. A rock is a rock. A planet is a planet. A neuron is a neuron. A brain is a brain. A brain cannot "partake of" a spider. A spider is a spider. My experiencing a spider cannot be a relation between a mysterious self and an extraneous spider. My experiencing a spider is an identity between the spider and what I am, experience. I am the spider; it's the only thing that has the right properties to be my experience.

Parks: The spider and simultaneously all the other object-experiences of your life. So, identity is constantly expanding.

Manzotti: Western thought has always sought to describe consciousness as identical with something, whether that be ideas, in the idealist view, or neural activity, according to the contemporary materialist position. When I propose an identity between consciousness and the world, I am following the same explanatory strategy scientists and philosophers have always adopted. I have simply settled on a new candidate for identity: not ideas, not neurons, but the world itself.

Parks: Then consciousness has always been, as they say, *hidden in plain sight.*

Manzotti: Exactly. We are the world that surrounds our body and the body itself as known; we are the objects we see, hear, smell, taste, touch. The rest is hot air.